Industrial Engineering

Industrial Engineering

Ajay Shrivastava

RANDOM PUBLICATIONS
NEW DELHI (INDIA)

Industrial Engineering

ISBN 978-93-5111-642-4

Published in 2015 in India by

RANDOM PUBLICATIONS

4376-A/4B, Gali Murari Lal, Ansari Road
New Delhi-110 002
Phone : +9111-43580356, 011-23289044, 011-43142548
e-mail: sales@randompublications.com,
info@randompublications.com, randomexports@gmail.com

Type Setting by : Friends Media, Delhi-110089
Printed at : Mehra Printers, Delhi-110 092

Preface

While the term originally applied to manufacturing, the use of "industrial" in "industrial engineering" can be somewhat misleading, since it has grown to encompass any methodical or quantitative approach to optimizing how a process, system, or organization operates. Some engineering universities and educational agencies around the world have changed the term "industrial" to broader terms such as "production" or "systems", leading to the typical extensions noted above. In fact, the primary U.S. professional organization for Industrial Engineers, the Institute of Industrial Engineers has been considering changing its name to something broader, although the latest vote among membership deemed this unnecessary for the time being. Industrial engineering is a branch of engineering which deals with the optimization of complex processes or systems. It is concerned with the development, improvement, implementation of integrated systems of people, money, knowledge, information, equipment, energy, materials, analysis and synthesis, as well as the mathematical, physical and social sciences together with the principles and methods of engineering design to specify, predict, and evaluate the results to be obtained from such systems or processes. While industrial engineering is a traditional and longstanding engineering discipline subject to professional engineering licensure in most jurisdictions, its underlying concepts overlap considerably with certain business-oriented disciplines such as operations management. Textile Engineering provides students with specific training in basic production fields, services and in relation to finished products.

I would like to thank my team for standing beside me throughout my career and writing this book. My special thanks go to "Random Publications" who have published the book.

– Ajay Shrivastava

Contents

1

Industrial Engineering in Textile

Industrial engineering is a branch of engineering which deals with the optimization of complex processes or systems. It is concerned with the development, improvement, implementation of integrated systems of people, money, knowledge, information, equipment, energy, materials, analysis and synthesis, as well as the mathematical, physical and social sciences together with the principles and methods of engineering design to specify, predict, and evaluate the results to be obtained from such systems or processes. While industrial engineering is a traditional and longstanding engineering discipline subject to (and eligible for) professional engineering licensure in most jurisdictions, its underlying concepts overlap considerably with certain business-oriented disciplines such as operations management.

Depending on the subspecialties involved, industrial engineering may also be known as, or overlap with, operations management, management science, operations research, systems engineering, management engineering, manufacturing engineering, ergonomics or human factors engineering, safety engineering, or others, depending on the viewpoint or motives of the user. For example, in health care, the engineers known as health management engineers or health systems engineers are, in essence, industrial engineers by another name.

While the term originally applied to manufacturing, the use of "industrial" in "industrial engineering" can be somewhat misleading, since it has grown to encompass any methodical or quantitative approach to optimizing how a process, system, or organization operates. Some engineering universities and educational agencies around the world have changed the term "industrial" to broader terms such as "production" or "systems", leading to the typical extensions noted above.

In fact, the primary U.S. professional organization for Industrial Engineers, the Institute of Industrial Engineers (IIE) has been considering changing its name to something broader (such as the Institute of Industrial and Systems Engineers), although the latest vote among membership deemed this unnecessary for the time being.

The various topics concerning industrial engineers include:

- Accounting: the measurement, processing and communication of financial information about economic entities
- Operations research, also known as management science: discipline that deals with the application of advanced analytical methods to help make better decisions
- Operations management: an area of management concerned with overseeing, designing, and controlling the process of production and redesigning business operations in the production of goods or services.
- Project management: is the process and activity of planning, organizing, motivating, and controlling resources, procedures and protocols to achieve specific goals in scientific or daily problems.
- Job design: the specification of contents, methods and relationship of jobs in order to satisfy technological and organizational requirements as well as the social and personal requirements of the job holder.
- Financial engineering: the application of technical methods, especially from mathematical finance and computational finance, in the practice of finance
- Engineering management: a specialized form of management that is concerned with the application of engineering principles to business practice
- Supply chain management: the management of the flow of goods. It includes the movement and storage of raw materials, work-in-process inventory, and finished goods from point of origin to point of consumption.
- Process engineering: design, operation, control, and optimization of chemical, physical, and biological processes.
- Systems engineering: an interdisciplinary field of engineering that focuses on how to design and manage complex engineering systems over their life cycles.
- Ergonomics: the practice of designing products, systems or processes to take proper account of the interaction between them and the people that use them.
- Safety engineering: an engineering discipline which assures that engineered systems provide acceptable levels of safety.
- Cost engineering: practice devoted to the management of project cost, involving such activities as cost- and control- estimating, which is cost control and cost forecasting, investment appraisal, and risk analysis.
- Value engineering: a systematic method to improve the "value" of goods or products and services by using an examination of function.
- Quality engineering: a way of preventing mistakes or defects in manufactured products and avoiding problems when delivering solutions or services to customers.

- Industrial plant configuration: sizing of necessary infrastructure used in support and maintenance of a given facility.
- Facility management: an interdisciplinary field devoted to the coordination of space, infrastructure, people and organization
- Engineering design process: formulation of a plan to help an engineer build a product with a specified performance goal.
- Logistics: the management of the flow of goods between the point of origin and the point of consumption in order to meet some requirements, of customers or corporations.

Traditionally, a major aspect of industrial engineering was planning the layouts of factories and designing assembly lines and other manufacturing paradigms. And now, in so-called lean manufacturing systems, industrial engineers work to eliminate wastes of time, money, materials, energy, and other resources. Examples of where industrial engineering might be used include flow process charting, process mapping, designing an assembly workstation, strategizing for various operational logistics, consulting as an efficiency expert, developing a new financial algorithm or loan system for a bank, streamlining operation and emergency room location or usage in a hospital, planning complex distribution schemes for materials or products, and shortening lines (or queues) at a bank, hospital, or a theme park. Modern industrial engineers typically use predetermined motion time system, computer simulation (especially discrete event simulation), along with extensive mathematical tools for modelling, such as mathematical optimization and queue theory, and computational methods for system analysis, evaluation, and optimization.

HISTORY

Efforts to apply science to the design of processes and of production systems were made by many people in the 18th and 19th centuries. They took some time to evolve and to be synthesized into disciplines that we would label with names such as industrial engineering, production engineering, or systems engineering. For example, precursors to industrial engineering included some aspects of military science; the quest to develop manufacturing using interchangeable parts; the development of the armory system of manufacturing; the work of Henri Fayol and colleagues (which grew into a larger movement called Fayolism); and the work of Frederick Winslow Taylor and colleagues (which grew into a larger movement called scientific management). In the late 19th century, such efforts began to inform consultancy and higher education. The idea of consulting with experts about process engineering naturally evolved into the idea of teaching the concepts as curriculum.

Industrial engineering courses were taught by multiple universities in Europe at the end of the 19th century, including in Germany, France, the United Kingdom, and Spain. In the United States, the first department of industrial and manufacturing engineering was established in 1909 at the Pennsylvania

State University. The first doctoral degree in industrial engineering was awarded in the 1930s by Cornell University.

In general it can be said that the foundations of industrial engineering as it looks today, began to be built in the twentieth century. The first half of the century was characterized by an emphasis on increasing efficiency and reducing industrial organizations their costs.

In 1909, Frederick Taylor published his theory of scientific management, which included accurate analysis of human labor, systematic definition of methods, tools and training for employees. Taylor dealt in time using timers, set standard times and managed to increase productivity while reducing labor costs and increasing the wages and salaries of the employees.

In 1912 Henry Laurence Gantt developed the Gantt chart which outlines actions the organization along with their relationships. This chart opens later form familiar to us today by Wallace Clark.

Assembly lines: moving car factory of Henry Ford (1913) accounted for a significant leap forward in the field. Ford reduced the assembly time of a car more than 700 hours to 1.5 hours. In addition, he was a pioneer of the economy of the capitalist welfare ("welfare capitalism") and the flag of providing financial incentives for employees to increase productivity.

Comprehensive quality management system (TQM) developed in the forties was gaining momentum after World War II and was part of the recovery of Japan after the war.

In 1960 to 1975, with the development of decision support systems in supply such as the MRP, you can emphasize the timing issue (inventory, production, compounding, transportation, etc.) of industrial organization. Israeli scientist Dr. Jacob Rubinovitz installed the CMMS programme developed in IAI and Control-Data (Israel) in 1976 in South Africa and worldwide.

In the seventies, with the penetration of Japanese management theories such as Kaizen and Kanban West, was transferred to highlight issues of quality, delivery time, and flexibility.

In the nineties, following the global industry globalization process, the emphasis was on supply chain management, and customer-oriented business process design. Theory of constraints developed by an Israeli scientist Eliyahu M. Goldratt (1985) is also a significant milestone in the field.

ENGINEERING DESIGN OF TEXTILES

OLD AND NEW DESIGN CULTURES

What is engineering design? And to what extent is it practised in the textile industry? Until about 100 years ago, engineering was an empirical skill. Bridges, buildings, machines, vehicles and all the other constructs needed by mankind were designed on the basis of experience and practical trials – with disastrous errors often leading to advances. Geometry linked to drawing was backed up by empirical rules of thumb on what loads could be carried and what might be needed to meet other performance requirements. The wealth of practical knowledge and skills developed over thousands of years enabled the production of textile materials to be an outstanding example of this form of engineering design. Experience and trial-and-error formed the design protocol; intuition led to advances. Point paper and weaving plans were the design tools. The flowering of textile research in research institutes and universities after the First World War led to an exploration of the academic science of fibres and textiles. In a commercial and industrial context, this led to advances through a growth of qualitative understanding, but the empirical design tradition remained as it was. Even the exploitation of the new manufactured fibres followed an empirical path of development.

The possible quantitative performance predictions of the academic work were not taken up by industry. The combination of strength and flexibility in unbonded fibre assemblies held together by frictional forces makes wonderful products. From ancient times, the needs for clothing of the proletariat and fashion for the aristocrats were provided by woven and knitted fabrics. A range of technical uses from sails and tents to conveyor belts and tyre fabrics was also met by textiles. The current *Extreme Textiles* exhibition at the Cooper-Hewitt National Design Museum in New York shows how this tradition has been exploited in new materials for space, medicine, sport, buildings, and transportation, as well as advances in more traditional applications such as ropes and nets. However a conversation with a manufacturer of materials for medical implants confirmed the preference for working through traditional practical trials – and eschewing mechanical modelling. In most other branches of engineering, design procedures changed.

The growth of the science of applied mechanics, starting with Galileo and Newton and augmented by giants like Euler and Hamilton as well as others who worked on the details, made quantitative predictions possible. In the education of mechanical and civil engineers, the design tools were provided in the first half of the 20th century by such classic works as Love's *Treatise on the Mathematical Theory of Elasticity*, Timoshenko's several books and, as a simpler text-book, Den Hartog's *Strength of Materials*, with Roark and Young's *Formulas for Stress and Strain* being a *vade mecum* of the equations that design engineers evaluated on their slide rules. Other scientific advances led to quantitative design in aerodynamic, electrical and chemical engineering. In the second half of the 20th century, computers replaced slide rules and computer-aided design (CAD) made more elaborate calculations and modelling possible. The application of this new form of quantitative engineering design, which replaced empiricism and augmented the qualitative insights, led to enhanced performance in traditional uses, such as bridges, and made possible the technological transformation of the 20th century. In most respects the changing design culture passed by the textile industry. There are some exceptions. The machinery spawned the academic subject of *mechanics of machines*, covering gears. levers and drives. The textile machinery industry has embraced CAD – but this is almost entirely in the machine actions independently of the fibres and yarns passing through the machines.

For example, a conversation with company engineers indicated that there was no attempt to model the movement of fibres in air-streams in the outstanding development of air-jet spinning by *Murata*; intuition and practical trials led to the commercial success. Modern engineering design is used for fibre-reinforced composites, which notably are included as *Extreme Textiles* in the exhibition, but these were developed in the mechanical engineering sector and have an affinity with rigid materials, subject only to small strains. Rope engineering is a particular example discussed below. In the application of science, there is a major difference between the two halves of the textile processing industry.

The chemical part, wet processing, has embraced science. The manager of a dyehouse or a finishing plant would regard himself as a chemist. Not so in the mechanical part. The manager of a spinning plant or a weaving shed would not regard himself as a member of the applied mechanics community, even though his work is overseeing mechanical actions and the mechanical performance of the textile products.

CONDITIONS FOR A CULTURAL CHANGE

Why has the textile industry not adopted a modern engineering design culture? There are several reasons. One is conservatism. Another is that it is not absolutely necessary. Bridges must be built so that they do not fall down; aircraft must fly. The structures are large; the costs are high; performance continually needs to be increased; the design process for new models is long. Quantitative engineering design is needed for safety and to optimise cost-benefit. In contrast to this, most textile products are small-scale; individual costs are low; failure in a trial is a nuisance but not a disaster; design times are short. There is a more fundamental reason.

The science of flexible fibre assemblies is an extremely challenging subject and is little related to the mainstream 20th century development of applied mechanics. As I wrote in 1979: *Textiles are solid materials, but little of direct relevance to textile behaviour will be found in any textbook on the mechanics of materials. Table lists some features which distinguish textile materials from those usually studied in engineering materials.*

In ordinary engineering, the development of discontinuities, of porosity, of buckling, of longrange displacement, of surface roughness, or of a soft elastic yielding under transverse pressure are often taken as signs of failure of the materials, and the need for mechanical analysis ceases with the onset of these phenomena: but in textiles their manifestation signals the value of these materials, and the beginning of the region where the mechanical analysis is of most interest. The convenient assumptions of small strains in an isotropic continuum are not applicable to flexible fibre assemblies, and fibre materials are not elastically governed by Hooke's Law with two elastic constants or ideal elastic-plastic, but are anisotropic and visco-elastic.

Table. Important Distinctions

common engineering materials	*intermediate categories*	*textile materials*
RIGID	flexible solid	FLEXIBLE
HOMOGENEOUS	sheets	DISCONTIUOUS
HARD		SOFT
IMPERVIOUS	soft loose	POROUS
SMOOTH	fillings of	TEXTURES
STRONG	no strength	STRONG

The above causes lead to another. There has been no creative interchange between those doing the basic research and engineers using it. The best academic work has opened up the fundamentals of the subject; the worst has concentrated on mathematical sophistication, which is of little practical value. Textile engineers have not found methodology that they could apply. There are several reasons why this is bound to change in the 21st century. The general development of IT culture means that young people entering industry expect to use computers to help solve their problems.

The wealth of practical experience, which was available when a textile technologist worked in the same area throughout a career, is dissipating with changing employee patterns.

The diversity of fibres and textile manufacturing methods is increasing greatly, opening up more choices and more problems. Textiles are finding uses in more demanding applications where the project engineers expect engineering data on the performance of the material. If the techniques are available, modelling predictions are cheaper than time-consuming pilot studies, and are welcomed by the engineering community. The fundamental reason is in Moore's Law.

The huge increase in computer power and reduction in cost means that the difficult problems can be tackled. The history of computing is that advances are first used by specialists and then taken up by the majority. Once started the changes occur rapidly. In the aesthetic aspects of the industry, designing for pattern and colour, CAD went from almost nowhere in 1975 to common practice by 2000. Although research has elucidated only a few of the problems in textile mechanics to a level for industrial performance prediction, the more urgent priority is to introduce programmes that are easy to use and truly help the engineer in industry.

CURRENT STATE OF THE ART

Yarns and Ropes

The mechanics of twisted continuous filament yarns had been effectively worked out by the 1960s. An energy method led to one integral and four algebraic nonlinear equations, which are easily solved numerically. The inputs are the fibre stress-strain curve and yarn linear density and twist. Although there are some approximations in the model, the predictions are good, except that the stresses at low strain are slightly less than expected because of some buckling of filaments at the centre of yarns. The reason for success is that twisted continuous filament yarns can be modelled by a well-defined geometry, which consists of superimposed layers of helices with constant twist period, and consequently there is a direct relation between yarn extension and the extension of the filaments.

In real yarns, the filaments migrate in radial position, but, over short lengths, the idealised model is a good approximation. The same methodology has been applied to the multiple twist levels of ropes. A hierarchical procedure is followed. Typically the sequence of twist levels is [yarn as produced] ! [rope yarn] ! [strand] ! [rope]. The inputs to the computer programme, *Fibre Rope Modeller* (FRM), are the dimensions and tensile properties of the component yarns and the twist inserted at each manufacturing stage. The predictions agree well with measured rope load-elongation properties, except that typically break loads of well-made ropes are about 10 per cent less than computed due to lack of optimum load-sharing. The geometric modelling enabled relative fibre motions in various slip modes to be computed and, applying the principle of virtual work, the internal transverse pressures between rope components to be determined. FRM is now being used by the leading UK rope manufacturer to design high-strength ropes, such as the 2000 tonne break-load polyester rope used to moor an oil-rig in 1400 meters of water in the Gulf of Mexico. This is one of the first examples of a textile design calculation being used in a modern engineering design sense. FRM was extended to compute responses under cyclic loading conditions.

The modelling takes account of fibre creep and hysteresis heating. Slip and pressure are determinants of inter-fibre abrasion. Buckling of some components at low rope tensions leads to axial compression fatigue; modelling was adapted from previous studies of buckling of heated pipelines. This part of FRM gives useful guidance but is limited in predictive power by lack of experimental values for rates of fibre abrasion and axial compression fatigue. Collaboration between marine engineering consultants, *Noble Denton*, and fibre rope consultants, *Tension Technology International* (TTI), supported by a consortium of rope makers, rope users and certifying authorities, led to the production of *Deepwater Fibre Moorings; an engineers' design guide*. The main criteria to be met in use are peak loads safely below rope break load, rig offsets less than an acceptable distance, and long enough life. Mooring analysis programmes existed to predict the response of steel wire and cable moorings. However, these moorings are controlled by the weight of the rope and changes in the catenary, whereas as fibre rope moorings are controlled by the tension due to the extension of a taut rope. In principle, tension is given by (modulus x extension). The problem is that viscoelasticity means that fibre modulus is not a well-defined property. It depends on the nature of the loading and the previous loading history.

There is the further complication that rope constructions tighten up under cyclic loading. The collaboration made it possible to specify test procedures to give a minimum value of *post-installation stiffness*, which related to offsets, and a maximum value for *storm stiffness*, which related to peak load. Limits were given for the number of cycles at low tension to avoid axial compression fatigue (e.g. <100,000 cycles at <10 per cent of break load for polyester; <2,000 cycles at <5 per cent of break load for aramids). A number of Joint Industry Projects, carried out by TTI and the National Engineering Laboratory, of cyclic loading on large test-rigs have provided data on moduli and effects of long-term testing (with the right products, no failure after millions of cycles). Polyester ropes have the best combination of properties for deepwater moorings and their long-term life is better than for steel ropes under similar conditions.

The combination of modelling and testing has given oil companies the confidence to moor oil-rigs in deep water. A typical installation would consist of 20 km of 50 Mtex polyester rope with break load of 2000 tonne and, in total, weighing 1000 tonnes. Collaboration between textile experts and applications engineers is the way forward for engineering design of textiles for demanding uses. Combinations of theoretical modelling and practical testing are needed to meet performance and safety requirements and to optimise cost-benefits. Experience in use must also be taken into account. Unexpected problems cannot be avoided. For example, in an early deep water mooring off the coast of Brazil, a rope section removed for testing showed a reduced break load due to minute crustaceans penetrating the rope and abrading the fibres. Once recognised, this problem could be eliminated by protecting the rope or substituting a steel section at the relevant depth, but it had not been anticipated despite extensive design studies.

Other yarns

The treatment of continuous filament yarns can be modified to take account of slip at fibre ends in compact ring-twisted staple fibre yarns. An elaborate theoretical analysis, which took account of fibre migration but was limited to small strains, was reduced to a semi-empirical equation: yarn strength or modulus/fibre strength or modulus

$$= \cos^2\alpha\,[\,1 - (2\ \mathrm{cosec}\alpha/3L)\,(a\,Q/2\,\mu)^{1/2}\,] = \cos^2\alpha\,(1 - K\ \mathrm{cosec}\alpha)$$

where ˜ = surface twist angle, L = fibre length, a = fibre radius, Q = migration period and ˜ = coefficient of friction.

Now it would be possible to apply the same principles to an individual-fibre computational model to give improved predictions. The major limitation would be to define the migration paths of the fibres. A full theoretical computation would require treatment of the more difficult problem of modelling the process of yarn formation.

An alternative design methodology would be to be to create a data-base of values of the empirical factor K and use a neural network or other soft procedure to give values for particular cases. The several forms of open-end spun yarns require modelling in terms of their structures. Bulky wool yarns have proved more difficult to model, because the structure changes as the fibres move closer together. Modelling of false-twist textured yarns, with alternating fibre helices and pig-tails, and air-jet textured yarns. with projecting loops, needs to be advanced to take account of modern computer power.

WOVEN FABRICS

The mechanics of woven fabrics raises more difficulties. One problem is that the zero-stress state is not well defined. Agreement is needed on the definition of a reference state under low biaxial loading. Another difficulty is that there is no direct geometrical relation between the fabric deformation and the component yarn deformations, since the latter depend on a balance between yarn extension, yarn bending and yarn flattening. One consequence of this is that fabric geometry as specified will not in general be in equilibrium. The first stage of a mechanical analysis is to determine the zero-stress state. In practice this is complicated by the ways in which fabrics can become set in different states as a result of relaxation.

Furthermore, the yarns go through different states as they go from contact regions at crossovers to free regions between crossovers, with more complicated forms in nonplain weaves. The majority of the many papers on woven fabric mechanics have been based on equilibrium of forces and moments. Most have been limited to plain weaves and simplifying assumptions, such as Kawabata's saw-tooth model, have been necessary. In my opinion, this approach will not lead to engineering design methods. The alternative is the energy-based approach of Hearle and Shanahan.

This is now being more strongly developed in a graphical computing mode in a research project at the University of Manchester. The fabric geometry is flexibly formulated by defining yarn paths by spline-fitting through specified points and by defining yarn cross-section by the lengths of a set of radial vectors. The mechanics is treated by minimisation of [yarn extension energy + yarn bending energy + yarn flattening energy] through adjustments of the geometry when the fabric is held to given dimensions. Determination of strain energy at small increments of deformation means that tensile forces can be found from conservation of energy. Yarn extension energy is well defined. Yarn bending energy is fairly well understood. The neglected feature of yarn flattening energy, which depends on change of volume and shape, needs more research. Progress can be helped in other ways. One is to concentrate on control points. Three primary control points, given by the fabric repeat, are needed to specify the fabric's biaxial deformation. Additional primary control points are needed for fabric shearing and bending. For a plain weave, the primary control points also define the internal displacements, but additional secondary control points are needed for non-plain weaves. Another is to identify special cases. For monofilament and highly twisted yarns, yarn flattening can be ignored, except for a Poisson's ratio effect usually at constant volume. For very soft yarns, the free zone between crossovers disappears. Simplifications of this sort, although lacking in academic rigour and generality, offer better prospects of developing useful engineering design software.

OTHER FABRICS

In mechanical terms, braided fabrics are simply a variant of woven fabrics. Simple plain knits have been modelled by equilibrium of forces and moments with limitations similar to those for woven fabrics. The energy method should be developed for the various knit fabric structures. Modelling of bonded, needled and stitch-bonded nonwovens has followed the familiar pattern of aiding qualitative understanding but not quantitative engineering design predictions. Random networks could be modelled by individual fibre computations, but the problem, as with staple yarns, is to define yarn paths. In the absence of detailed process modelling, semi-empirical procedures are needed for engineering design.

OTHER MECHANICAL PROPERTIES

The above discussion has been primarily concerned with tensile properties, but there are others that are important for textile performance. The transverse compressibility of yarns has been mentioned in connection with flattening energy in fabrics. For fabrics, there are six fundamental directions to be modelled: in-plane, two tensile and one shear; out-of-plane, two bending and one twisting, the latter, which is related to bending on the bias, being frequently ignored. Solution of these problems covers fabric micromechanics and gives the constitutive relations for yarns and fabrics. The hierarchy is:

fibre properties + yarn structure
↓
yarn properties + fabric structure
↓
fabric properties

Macromechanics covers more complex deformations and modes of behaviour, which are relevant to a totality of engineering design. These include features such as pilling and bagging of fabrics, where modelling has been limited, and heat and moisture transmission, where modelling is advanced. Two examples illustrate ways to approach the problems. Many attempts at modelling of drape by finite-element programmes have been reported in the literature. However, these include unjustified simplifications and demand excessive computer time.

In my view, it is not a useful approach to apply software developed for other systems. Programmes adapted to the particular features of textiles are needed. Drape is just one manifestation of complex buckling of fabrics. I believe that progress will come from a fundamental study of the mechanics of buckling, elaborating the study of symmetrical threefold buckling of a linear elastic sheet to cover more realistic and complicated forms. Bulk and compressibility of fibre assemblies is another important problem. Here the pioneering individual-fibre computational modelling of Beil and Roberts shows the way forward. Again this is based on energy minimisation of the forms of fibre segments between contact points.

USEFUL SOFTWARE

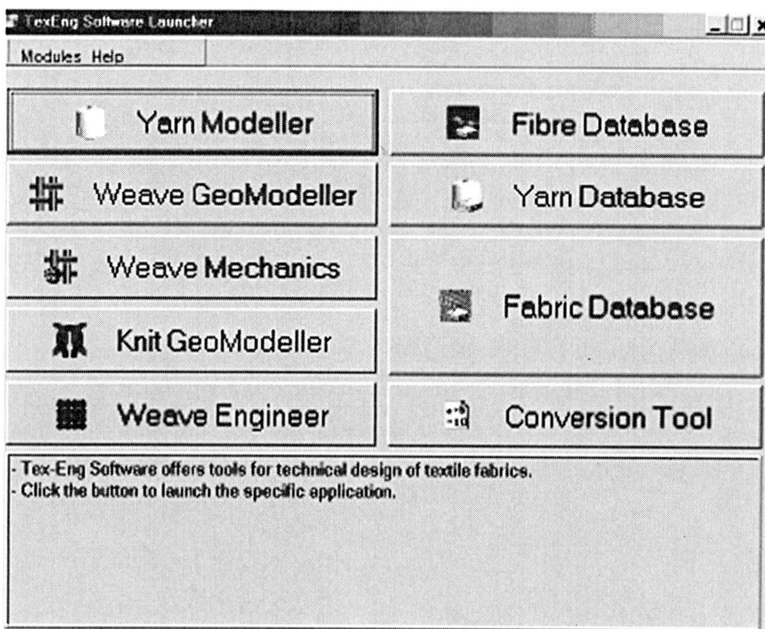

Fig. Entry screen for TexEng Programmes.

How can we make the increasing knowledge of the structural mechanics of textiles into engineering design procedures that will be used in industry? This is necessary, not only for its inherent value, but as a means of developing a creative interchange between academic research and practical utility. The critical need is for programmes that are easy to use and answer the questions that engineers and managers want answered. In TexEng Ltd, we offer programmes developed in the Textiles and Paper Group, School of Materials, University of Manchester in association with TTI Ltd in forms which run in a familiar WINDOWS mode on a PC. The entry screen is shown in Fig.

Yarn Modeller covers the established model for twisted continuous filament yarns and will include empirical rules for other yarns. Weave GeoModeller covers the structure of single-layer woven fabrics. Fig. 2 shows montages from several windows. A file is created in three stages. (1) Weave can be input by traditional point paper, mesh, which is easy for computer usage and more illustrative, or, for the mathematically inclined, by formula. (2) Yarn dimensions are input; circular, lenticular and racetrack cross-sections are current options. (3) Warp and weft spacings and one crimp value complete the fabric specification. If needed, conversion tool transforms input values for other parameters to those required for the programme.

Three-dimensional views of the fabric can be manipulated in various ways. For example, slice allows the viewer to see how pore sizes change through the thickness of the material, which is relevant to transmission properties. Weave GeoModeller is the basic programme that allows particular applications to be developed. For example, in one project in the University of Manchester, a filtration module is being developed. Knit GeoModeller is similar to Weave GeoModeller but is currently limited to plain, rib and purl single-layer and interlock rib double–layer weft knits.

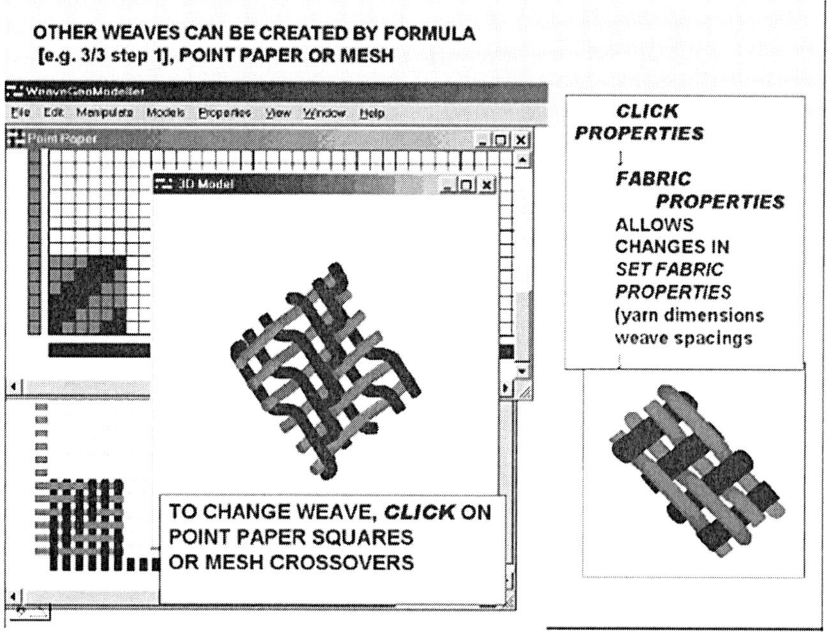

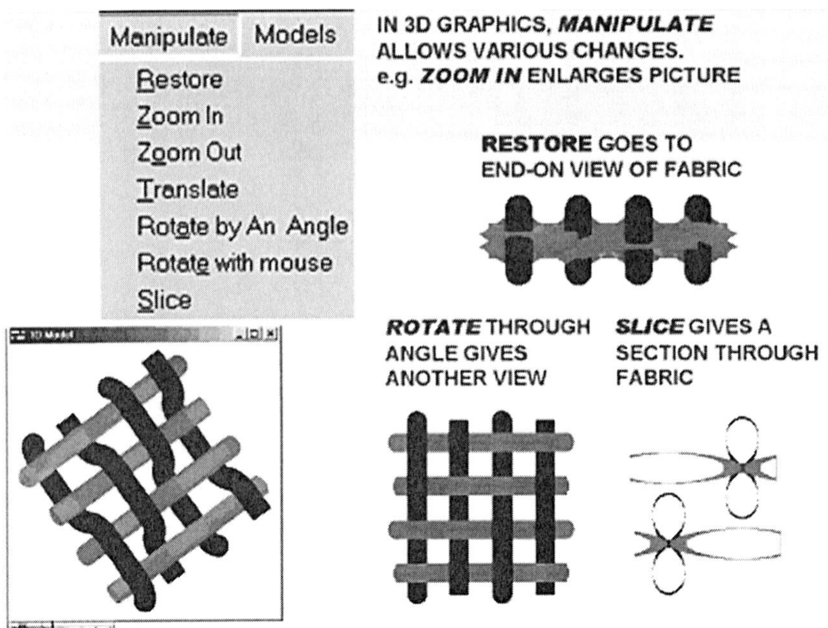

The currently available application in TexEng is Weave Mechanics, which is based on an energy method described by Sagar. The additional inputs are yarn tensile properties, specified by break load, break extension and shape of stress-strain curve through polynomial coefficients, bending stiffness and flattening stiffness.

The predictions agree well with experimental data on plain-weave cotton fabrics from Kawabata et al and monofilament nylon fabrics from Dastoor et al. However, as indicated above, more theoretical research and validation is needed for more complicated woven fabric problems. Fibre Data-Base and Yarn Data-Base, which can also store results from Yarn Modeller, are sources of inputs to the modelling programmes.

Fabric Data-Base stores the files of models that have been run. It can also be used to store experimental data on fabrics; ways of using this data to give empirical predictions will be developed. Conversion Tool derives from a pioneering development of a question-answer programme by Konopasek. Formulation of a great many textile problems involves a multiplicity of relevant variables, a few of which are independent and more are dependent. However the choice of independent variables depends on circumstances. Fibre dimensions are a trivial example.

Three alternatives are: given fibre diameter (˜m) and density (g/cm3), find linear density (tex); or, given linear density and density, find diameter; or, improbably, given linear density and diameter, find density. With woven fabrics, as illustrated in Fig. 3, the number of useful quantities is much larger: thread spacings or the reciprocals, ends per unit length, modular lengths, crimp values in various forms, weave angles, cover factors, fabric weight, etc for both warp and weft. The programme has a facility for switching between the variety of units that may be used. Other features, such as costings, are included. The programme can also cover empirical relations, such as: Weave Engineer is a programme developed by Xiaogang Chen and Isaac Porat for 3D fabrics, which are important for composite preforms. It covers the weave specification and a view of the topology of the fabric, as illustrated in Fig. 4. The information can be transmitted to the electronic Jacquards or other control means of weaving machines; this part of the programme can also be used with Weave GeoModeller for single-layer fabrics.

APPLICATION OF NANOTECHNOLOGY IN TEXTILE ENGINEERING

The fast development and changes in life style has attracted peoples towards a more comfort and luxurious life. People are moving towards small, safer, cheaper and fast working products which not only reduces the work load but also help them to carry out their works at a much greater pace with minimum efforts. There have been development of gadgets that are much smaller in size like micro-chip, nano capsules, carbon tubes, memory cards, pen drives etc which reduces the problems of transport, and storage and are also much faster and reliable by which we can carry out more of our work in less time. In the formation and development of such products nanotechnology plays a very important and vital role. The term nanotechnology (sometimes shortened to "nanotech") comes from nanometer – a unit of measure of one billionth of a meter of length. The concept of Nanotechnology was given by Nobel Laureate Physicist

Richard Feynman, in 1959. Nanotechnology is defined as the understanding, manipulation, and control of matter at the length scale on nanometer, such that the physical, chemical, and biological properties of materials (individual atoms, molecules and bulk matter) can be engineered, synthesized or altered to develop the next generations of improved materials, devices, structures, and systems. Generally, nanotechnology deals with structures that are sized between 1 to 100 nm in at least one dimension and involves developing materials or devices possessing dimension within that size. Nanotechnology creates structure that have excellent properties by controlling atoms and molecules, functional materials, devices and systems on the nanometer scale by involving precise placement of individual atoms.

Although nanotechnology is a relatively recent development in scientific research, the development of its central concepts happened over a longer period of time. The emergence of nanotechnology in the 1980s was caused by the convergence of experimental advances. The early 2000s also saw the beginnings of commercial applications of nanotechnology, although these were limited to bulk applications of nanomaterials, such as the silver nano platform for using silver nanoparticles as an antibacterial agent, nanoparticles-based transparent sunscreens, and carbon nanotubes for stain-resistant textiles.

Throughout history, the textiles sectors have been used worldwide in a wide range of consumer applications. Natural fibres, such as cotton, silk, and wool, along with synthetic fibres, such as polyester and nylon, continue to be the most widely used fibres for apparel manufacturing. Synthetic fibres are mostly suitable for domestic and industrial applications, such as carpets, tents, tires, ropes, belts, cleaning cloths, and medical products. Natural and synthetic fibres generally have different characteristics, which make them ideally suitable mainly for apparel.

Depending on the end-use application, some of those characteristics may be good, while the others may not be as good to contribute to the desired performance of the end product. As stated previously, nanotechnology brings the possibility of combining the merits of natural and synthetic fibres, such that advanced fabrics that complement the desirable attributes of each constituent fibre can be produced.

In the last decade, the advent of nanotechnology has spurred significant developments and innovations in this field of textile technology. By using nanotechnology, there have been developments of several fabric treatments to achieve certain enhanced fabric attributes, such as superior durability, softness, tear strength, abrasion resistance, durable-press and wrinkle-resistance. The (nano)-treated core component of a core-wrap bi-component fabric provides high strength, permanent anti-static behaviour, and durability, while the traditionally-treated wrap component of the fabric provides desirable softness, comfort, and aesthetic characteristics.

The use of nanotechnology in the textile industry made it multifunctional which can produce fibres with variable functions and applications such as UV protection, antiodour, antimicrobial etc. In many cases smaller amounts of the additive are required, for the saving on resources. The success of Nanotechnology and its potential applications in textiles lies in various fields where new methods are combined with multifunctional textile systems, durable etc. without affecting the inherent properties of the textiles including softness, flexibility, washbility etc. Keeping the above factors in consideration, the present review highlighted the use of nanotechnology in textile industry and textile engineering, the types and methods of preparation of different nano composites used in textiles.

BASIC NANOTECHNOLOGY

Two main approaches are used in nanotechnology that is, the bottom up approach and the top-down approach. In case of the "bottom-up" approach, the different type of materials and the instruments are made up from different types of molecular components which combine themselves by chemical ways basing on the mechanism of molecular recognition. In case of the "top-down" approaches, various nano-objects are made from various types of components without atomic-level control. Materials reduced to the nanoscale can show different properties compared to what they exhibit on a macro scale, enabling unique applications.

The basic premise is that properties can dramatically change when a substance's size is reduced to the nanometer range. For instance, ceramics which are normally brittle can be deformable when their size is reduced, opaque substances become transparent (copper); stable materials turn combustible (aluminum); insoluble materials become soluble (gold).

Nanoparticles can be prepared from a variety of materials such as proteins, polysaccharides and synthetic polymers. The selection of materials in mainly depended on factors like size of the nanoparticles required, inherent properties such as aqueous, solubility and stability, surface characteristics i.e. charge and permeability, degree of biodegradability, biocompatibility and toxicity, release of the desired product, antigenicity of the final product etc. Polymeric nanoparticles have been prepared most frequently be three methods (1) Dispersion of the performed polymers; (2) Polymerization of the monomers and (3) Ionic gelation of the hydrophobic or hydrophilic polymers. However, techniques like supercritical fluid technology and particle repulsion in non-wetting templets (PRINT) have been also used in modern days.

NANOTECHNOLOGY IN TEXTILE INDUSTRY AND TEXTILE ENGINEERING

Of the many applications of nanotechnology, textile industry has been currently added as one of the most benefited sector. Application of nanotechnology in textile industry has tremendously increased the durability of fabrics, increase its comfortness, hygienic properties and have also reduces its production cost. Nanotechnology also offers many advantages as compare to the conventional process in term of economy, energy saving, eco-friendliness, control release of substances, packaging, separating and storing materials on a microscopic scale for later use and release under control condition. The unique and new properties of nanotechnology have attracted scientists and researchers to the textile industry and hence the use of nanotechnology in the textile industry has increased rapidly.

This may be due to the reason that textile technology is one of the best areas for development of nanotechnology. The textile fabrics provide best suitable substrates where a large surface area is present for a given weight or a given volume of fabric. The synergy between nanotechnology and textile industry uses this property of large interfacial area and a drastic change in energetic is experienced by various macromolecules or super molecules in the vicinity of a fibre when changing from wet state to a dry state.

The application of nanoparticles to textile materials have been the objective of several studies aimed at producing finished fabrics with different functional performances. Nanoparticles can provide high durability for treated fabrics as they posses large surface area and high surface energy that ensure better affinity for fabrics and led to an increase in durability of the desired textile function. The particle size also plays a primary role in determining their adhesion to the fibres. It is reasonable to except that the largest particle agglomerates will be easily removed from the fibre surface, while the smallest particle will penetrate deeper and adhere strongly into the fabric matrix. Thus, decreasing the size of particles to nano-scale dimensional, fundamentally changes the properties of the material and indeed the entire substance.

A whole variety of novel nanotech textiles are already on the market at this moment. Areas where nanotech enhanced textiles are already seeing some applications include sporting industry, skincare, space technology and clothing as well as materials technology for better protection in extreme environments. The use of nanotechnology allows textiles to become multifunctional and produce fabrics with special functions, including antibacterial, UV-protection, easy-clean, water- and stain repellent and anti-odour.

Types of nanomaterials

Nanocomposite fibres

A composite is a material that combines one or more separate components. Composites are designed to exhibit the best properties of each component. A large variety of systems combining one, two and three dimensional materials with amorphous materials mixed at the nanometer scale. Nanostructure composite fibres are intensively used in automotive, aerospace and military applications. Nanocomposite fibres are produced by dispersing nanosize fillers into a fibre matrix. Due to their large surface area and high aspect ratio, nanofillers interact with polymer chain movement and thus reduce the chain mobility of the system. Being evenly distributed in polymer matrices, nanoparticles can carry load and increase the toughness and abrasion resistance. Most of the nanocomposite fibres use fillers such as nanosilicates, metal oxide nanoparticles, graphite nanofibres (GNF) as well as single-wall and multi-wall carbon nanotubes (CNT). Some novel CNT reinforced polymer composite materials have been developed, which can be used for developing multifunctional textiles having superior strength, toughness, lightweight, and high electrical conductivity.

Carbon nanofibres and carbon nanoparticles

Carbon nanofibres and carbon black nanoparticles are among the most commonly used nanosize filling materials. Nanofibres can be defined as fibres with a diameter of less than 1 mm or 1000 nm and are characterized as having a high surface area to volume ratio and a small pore size in fabric form. Carbon nanofibres can effectively increase the tensile strength of composite fibres due to its high aspect ratio, while carbon black nanoparticles can improve their abrasion resistance and toughness. Several fibre-forming polymers used as matrices have been investigated including polyester, nylon and polyethylene with the weight of the filler from 5 to 20 per cent.

There are numerous applications in which nanofibres could be suited. The high surface area to volume ratio and small pore size allows viruses and spore-forming bacterium such as *anthrax* to be trapped. Filtration devices and wound dressings are just some of the applications in which nanofibres could be utilized. Researchers are investigating textile materials made from nanofibres which can act as a filter for pathogens (bacteria, viruses), toxic gasses, or poisonous or harmful substances in the air. Medical staff, fire fighters, the emergency services or military personnel could all benefit from protective garments made fromnanofibres materials.

Clay nanoparticles

Clay nanoparticles are resistant to heat, chemicals and electricity, and have the ability to block UV light. Incorporating clay nanoparticles into a textile can result in a fabric with improved tensile strength, tensile modulus, flexural strength and flexural modulus. Nanocomposite fibres which utilize clay nanoparticles can be engineered to be flame, UV light resistant and anti-corrosive. Although there have been a number of flame retardant finishes available since the 1970's, the emission of toxic gasses when set ablaze make them somewhat hazardous. Clay nanoparticles have been incorporated into nylon to impart flame retardant characteristics to the textile without the emission of toxic gas. The addition of clay nanoparticles has made polypropylene dyeable. Metal oxide nanoparticles of TiO2, Al2O3, ZnO and MgO exhibit photocatalytic ability, electrical conductivity, UV absorption and photo-oxidizing capacity against chemical and biological species. The main research efforts involving the use of nanoparticles of metal oxides have been focused on antimicrobial, self-decontaminating and UV blocking applications for both military protection gears and civilian health products. Nylon fibres filled with ZnO nanoparticles can provide UV shielding function and reduce static electricity on nylon fibres. A composite fibre with nanoparticle of TiO2 or MgO can provide self-sterilizing function.

Carbon nanotubes

Carbon Nanotube is a tubular form of carbon with diameter as small as nanometer (nm). A carbon nanotube (CNT) is configurationally equivalent to a two dimensional graphene sheet rolled into a tube. They can be metallic or semiconducting, depending on chirality.CNT are one of the most promising materials due to their high strength and high electrical conductivity. CNT consists of tiny shell(s) of graphite rolled up into a cylinder(s). CNT exhibit 100 times the tensile strength of steel at one-sixth weight, thermal conductivity better than all but the purest diamond, and an electrical conductivity similar to copper, but with the ability to carry much higher currents. The potential applications of CNTs include conductive and high-strength composite fibres, energy storage and energy conversion devices, sensors, and field emission displays. Possible applications include screen displays, sensors, aircraft structures, explosion-proof blankets and electromagnetic shielding. The composite fibres have potential applications in safety harnesses, explosion-proof blankets, and electromagnetic shielding applications. Continuing research activities on CNT fibres involve study of different fibre polymer matrices such as polymethylmethacrylate (PMMA) and polyacrylonitrile (PNA) as well as CNT dispersion and orientation in polymers.

Nanocellular foam structure

Polymeric materials with nanosize porosity exhibit lightweight, good thermal insulation, as well as high cracking resistance at high temperature without sacrifices in mechanical strength. By choosing the pretreatment condition to the fibre, the transverse mechanical properties of the composite can be also enhanced through the molecular diffusion across the interface between the fibre and the matrix. The nanocomposites clearly surpass the mechanical properties of most comparable cellulosic materials, their greatest advantage being the fact that they are fully bio-based and biodegradable, but also of relatively high strength. A potential application of cellular structure is to encapsulate functional compounds such as pesticides and drugs inside of the nanosizecells. One of the approaches to fabricate nanocellular fibres is to make use of a thermodynamic instability during supercritical carbon dioxide extrusion and reduce the size of the cellular fibres that can be used as high-performance composite fibres as well as for sporting and aerospace materials.

Properties of Nano Textile Fibres

Water Repellence

The water-repellent property of fabric created by nano-whiskers, which are hydrocarbons and 1/1000 of the size of a typical cotton fibre, when added to the fabric create a peach fuzz effect without lowering the strength of cotton. The spaces between the whiskers on the fabric are smaller than the typical drop of water, but still larger than water molecules; water thus, remains on the top of the whiskers and above the surface of the fabric. However, liquid can still pass through the fabric, if pressure is applied to it.

Nanosphere impregnation involving a three-dimensional surface structure with gel forming additives which repel water and prevent dirt particles from attaching themselves are also used. Once water droplets fall onto them, water droplets bead up and, if the surface slopes slightly, will roll off. As a result, the surfaces stay dry even during a heavy shower. Furthermore, the droplets pick up small particles of dirt as they roll, and so the leaves of the lotus plant keep clean even during light rain. By altering the micro and nano-scale surface features on a fabric surface, a more robust control of wetting behaviour can be attained. It has been demonstrated that by combining the nanoparticles of hydroxylapatite, TiO2, ZnO and Fe7O3 with other organic and inorganic substances, the audio frequency plasma of fluorocarbon chemical was applied to deposit a nanoparticulate hydrophobic film onto a cotton fabric surface to improve its water repellent property.

This sort of surface engineering, which is capable of replicating hydrophobic behaviour, can be utilized in developing special chemical finishes for producing water-and/or stain- resistant fabrics while complementing the other desirable fabric attributes, such as breathability, softness and comfort.The surfaces of the textile fabrics can be appreciably modified to achieve considerably greater abrasion resistance, ultraviolet (UV) resistance, electromagnetic and infrared protection properties.

UV-protection

Inorganic UV blockers are more preferable to organic UV blockers as they are non-toxic and chemically stable under exposure to both high temperatures and UV. Inorganic UV blockers are usually certain semiconductor oxides such as TiO2, ZnO, SiO2 and Al2O3. Among these semiconductor oxides, titanium dioxide (TiO2) and zinc oxide (ZnO) are commonly used. It was determined that nano-sized titanium dioxide and zinc oxide are more efficient at absorbing and scattering UV radiation than the conventional size, and are thus better to provide protection against UV rays . This is due to the fact that nano-particles have a larger surface area per unit mass and volume than the conventional materials, leading to the increase of the effectiveness of blocking UV radiation. Various researchers have worked on the application of UV blocking treatment to fabric using nanotechnology.

UV blocking treatment for cotton fabrics are developed using the sol-gel method. A thin layer of titanium dioxide is formed on the surface of the treated cotton fabric which provides excellent UV protection; the effect can be maintained after 50 home launderings. Apart from titanium dioxide, zinc oxide nano rods of 10 to 50 nm in length are also applied to cotton fabric to provide UV protection. According to the studies on the UV blocking effect, the fabric treated with zinc oxide nanorods were found to have demonstrated an excellent UV protective factor (UPF) rating.

This effect can be further enhanced by using a different procedure for the application of nanoparticles on the fabric surface. When the process of padding is used for applying the nanoparticles on to the fabric, the nanoparticles get applied not only on the surface alone but also penetrates into the interstices of the yarns and the fabric, i.e. some portion of the nanoparticles get penetrate into the fabric structure. Such Nanoparticles which do not stay on the surface may not be very effective in shielding the UV rays. It is worthwhile that only the right (face) side of the fabric gets exposed to the rays and therefore, this surface alone needs to be covered with the nanoparticles for better UV protection. Spraying (using compressed air and spray gun) the fabric surface with the nanoparticles can be an alternate method of applying the nanoparticles.

Antimicrobial

Although many antimicrobial agents are already in used for textile, the major classes of antimicrobial for textile include organo-silicones, organo-metallics, phenols and quaternary ammonium salts. The bis- phenolic compounds exhibits a broad spectrum of antimicrobial activity. For imparting antibacterial properties, nano-sized silver, titanium dioxide,zinc oxide, triclosan and chitosan are used.

Nano-silver particles have an extremely large relative surface area, thus increasing their contact with bacteria or fungi and vastly improving their bactericidal and fungicidal effectiveness. Nano-silver is very reactive with protein and shows antimicrobial properties at concentrations as low as 0.0003 to 0.0005 per cent.

When contacting bacteria and fungi, it will adversely affect cellular metabolism and inhibits cell growth. It also suppresses respiration, the basal metabolism of the electron transfer system, and the transport of the substrate into the microbial cell membrane. Furthermore, it inhibits the multiplication and growth of those bacteria and fungi which cause infection, odour, itchiness and sores. Some synthetic antimicrobial nano particles which are used in textiles are as follows. Triclosan, a chlorinated bis- phenol, is a synthetic, non-ionic and broad spectrum antimicrobial agent possessing mostly antibacterial alone with some antifungal and antiviral properties. Chitosan, a natural biopolymer, is effectively used as antibacterial, antifungal, antiviral, non-allergic and biocompatible. ZnO nanoparticles have been widely used for their antibacterial and UV-blocking properties.

Antistatic

An antistatic agent is a compound used for treatment of materials or their surfaces in order to reduce or eliminate buildup of static electricity generally caused by the triboelectric effect. The molecules of an antistatic agent often have both hydrophilic and hydrophobic areas, similar to those of a surfactant; the hydrophobic side interacts with the surface of the material, while the hydrophilic side interacts with the air moisture and binds the water molecules. As synthetic fibres provide poor anti-static properties, research work concerning the improvement of the anti-static properties of textiles by using nanotechnology has been at large.

It was determined that nano-sized particles like titanium dioxide, zinc oxide whiskers, nanoantimony-doped tin oxide (ATO) and silanenanosol could impart anti-static properties to synthetic fibres. Such material helps to effectively dissipate the static charge which is accumulated on the fabric. On the other hand, silanenanosol improves anti-static properties, as the silica gel particles on fibre absorb water and moisture in the air by amino and hydroxyl groups and bound water. Electrically conductive nano-particles are durably anchored in the fibrils of the membrane of teflon, creating an electrically conductive network that prevents

the formation of isolated chargeable areas and voltage peaks commonly found in conventional anti-static materials. This method can overcome the limitation of conventional methods, which is that the anti-static agent is easily washed off after a few laundry cycles.

Wrinkle resistance

To impart wrinkle resistance to fabric, resin is commonly used in conventional methods. However, there are limitations to applying resin, including a decrease in the tensile strength of fibre, abrasion resistance, water absorbency and dye-ability, as well as breathability. To overcome the limitations of using resin, some researchers employed nano-titanium dioxide and nano-silica to improve the wrinkle resistance of cotton and silk respectively. Nano-titanium dioxide employed with carboxylic acid as a catalyst under UV irradiation to catalyses the cross-linking reaction between the cellulose molecule and the acid. On the other hand, nano-silica when applied with maleic anhydride as a catalyst could successfully improve the wrinkle resistance of silk. More over the wrinkle recovery of the fabrics can also be improved to a great extent by imparting techniques like padding and exhaustion beside the use of nano-materials to the fabrics.

Studies also have suggest that treatment of fabrics with microwaves are more wrinkle resistant as comparable to oven curing, because it generates higher frequency and volumetric heating which minimizes the damage from over drying.

Nanoparticles in textiles finishing

Fabric treated with nanoparticles of TiO2 and MgO replaces fabrics with active carbon, previously used as chemical and biological protective materials. The photocatalytic activity of TiO2 and MgO nanoparticles can break harmful and toxic chemicals and biological agents. These nanoparticles can be pre-engineered to adhere to textile substrates via spray coating or electrostatic methods. Textiles with nanoparticles finishing are used to convert fabrics into sensor-based materials which has numerous applications. If nanocrystalline epiezoceramic particles are incorporated intofabrics, the finished fabric can convert exerted mechanical forces into electrical signals enabling the monitoring of bodily functions such as heart rhythm and pulse if they are worn next to skin.

Fabric finishing by using nanotechnology

Finishing of textile fabrics made of natural and synthetic fibres to achieve desirable surface texture, colour and other special aesthetic and functional properties, has been a primary focus in textile manufacturing industries. In the last decade, the advent of nanotechnology has spurred significant developments and innovations in this field of textile technology. Fabric finishing has taken new routes and demonstrated a great potential for significant improvements by applications of nanotechnology.

The developments in the areas of surface engineering and fabric finishing have been highlighted in several researches. There are many ways in which the surface properties of a fabric can be manipulated and enhanced, by implementing appropriate surface finishing, coating, and/or altering techniques, using nanotechnology. A few representative applications of fabric finishing using nanotechnology are schematically displayed in the Figure.

Nanotechnology provides plenty of efficient tools and techniques to produce desirable fabric attributes, mainly by engineering modifications of the fabric surface. For example, the prevention of fluid wetting towards the development of water or stain-resistant fabrics has always been of great concerning textile manufacturing.

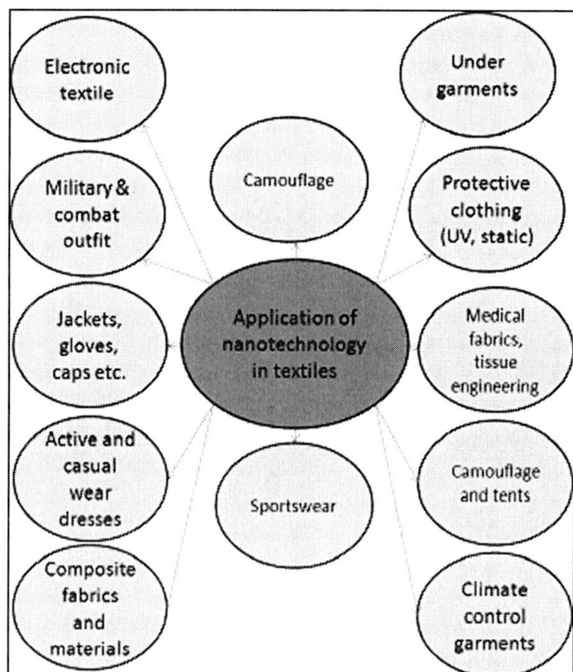

Fig. Some representative applications nanotechnology in textiles.

The basic principles and theoretical background of "fluid-fabric" surface interaction are well described in recent manuscript. It has been demonstrated that by altering the micro and nano-scale surface features on a fabric surface, a more robust control of wetting behaviour can be attained. The alteration in the fabric's surface properties enables to exhibit the "Lotus-Effect," which demonstrates the natural hydrophobic behaviour of a leaf surface. This sort of surface engineering, which is capable of replicating hydrophobic behaviour, can be utilized in developing special chemical finishes for producing water and/or stain- resistant fabrics.

In recent years, several attempts have been made by researchers and industries to utilize similar concepts of surface-engineered modifications through nanotechnology to develop high performance textile and smart textile. The concept of surface engineering and nanotextile develops hydrophobic fabric surfaces that are capable of repelling liquids and resisting stains, while complementing the other desirable fabric attributes, such as breathability, softness, and comfort.

CHALLENGES OF THE TEXTILE AND CLOTHING INDUSTRY AND TEXTILE EDUCATION

In 1989, at the time of the change of the political system in Hungary the textile and clothing industry lost its traditional markets faster than any other sector. Former state companies were wound up, with their places being taken by a plethora of small businesses. In the course of privatization the sector lost a huge amount of capital measured both in terms of means of production and in terms of human resources.

The privatized and restructured industry began to consolidate towards the end of the 1990s, and then started to grow, even if this growth started off only from a very low level. The number of enterprises increased rapidly in the course of the privatisation process until 1997. Since then the situation has been stabilized and the number of companies is even on the increase. State ownership had disappeared by 2000 almost entirely. However, after just a few years import competition began to intensify enormously, a phenomenon that was paralleled by an increasing tendency in the textile trade towards dubious business practices, primarily customs fraud and the avoidance of VAT. Immediately after the turn of the millennium market challenges brought about new forms in quality terms. The global division of labour strengthened, as did the move by manufacturers to relocate their activities to countries operating with lower costs

CURRENT SITUATION OF THE SECTOR

The sector is largely made up of small- and medium-size enterprises, although they are also attended by a considerable number of sole proprietorships. Besides companies a great many non corporate enterprises operate. These are registered as sole entrepreneurs and usually employ 1-5 people. Their number is decreasing; those of them who are successful become corporate enterprises. The number of registered enterprises approaches 4,500, although it is true to say that the number of functioning businesses is somewhat smaller than this. The sector currently employs around 50,000 people, the vast majority being female labour. There is a strong demand there

- to stop, stabilize and renew the Hungarian textile and clothing industry, with enhanced innovation and through changing the product structure
- to enlarge the number of competitive companies in the sector to save the actual level of employment
- to enhance the prestige of the sector

There are several factors to mention which negatively influence the development of the Hungarian textile and clothing industry:

- globalization - free trade, cheap import
- relative high labour costs
- increase in competition - countries with high productivity levels and low wages
- relocation of production to countries of Asia, Oceania, India and South America
- extremely unfavourable wage relations
- lack of state support
- low capital for the development of own products with high added value

The dramatic shrinkage of former capacities to produce hemp and flax in Hungary results in a danger, that more than 120 years of experience in production and processing will be lost. The knowledge in development of a drug-free industrial hemp plant and the "cottonization" of hemp is still available and could be used for the production of hemp-based green products especially in technical applications.

This could contribute to the development of the Southern-Plains region of Hungary and to building cross-border cooperation. 1-2 per cent of the Hungarian corn production (infected or waste for burning) could be used for the production of polylactid acid to produce polylactid fibre (PLA). The textile sector has become more innovative in the past years, approached new markets and applications for their products. Nowadays about 50 per cent of textile products are used for technical purposes.

Sub-sectors/technologies/products with potential from an export aspect:

- work and protective clothing
- technical textiles
- health care textiles
- home and household textiles
- other fashion items
- car carpets

Enterprises involved in the manufacture of technical textile products are relatively large, but mostly SME's with their own R andD strategies, to pursue continuous product development. Businesses active in this sub-sector are proving themselves increasingly successful not only on the Hungarian market, but on the wider European market, too.

SUSTAINABLE APPROACH - A CHALLENGE FOR THE MARKET PLAYERS OF THE TEXTILE AND CLOTHING SECTOR

Due to the increasing demand and to the higher profit resulting from the sale of products with environmental benefit, a new trend is gaining ground, green marketing. It can be a very powerful marketing strategy though when it's done properly. The trend results in the change of consumer behaviour and the traditional production processes. The strengthening of consumer awareness supports the protection of the environment buying more green products.

But sustainability means more than using environmentally-friendly materials, it is a strategy to save the environment, to reduce material and energy consumption, to be cost-effective in the production process and during the usage, equally considering environmental, social and economic aspects. It need a holistic approach, in which environmental, economic and social aspects are equal. There are three ways of managing the sustainable approach in the fashion sector:

- Product design and fabric/raw material selection (choosing more ecofriendly products and processes, (re-) design, re-examining markets, reusing waste and making well-designed and efficiently manufactured products)
- selecting retailers (choosing suppliers with a credible certificate, environmental policy)
- informing the consumer (seminars, consumer organisations)

There is a need for opening a dialogue between the brands and retailers and their supply chains to broaden integration of sustainability as a business strategy. The demand to know more about textile ecology is increasing, more and more events and organizations (Ethical Fashion Forum, the RITE Group, Textile Exchange, Made By, etc.) offer information on topics like green textiles, sustainable textile processing, eco-indexing, traceability and transparency, organic fibre production, and environmental footprint are discussed worldwide at many forums and conferences. A number of seminars are offered to provide an open and neutral ground to discuss most burning environmental issues, latest innovations, challenges, and the best practices on developing sustainable fashion production..

Raw materials for eco-fashion

Designers for eco-fashion prefer to use natural fibres or recycled materials. Even conscious consumers think that eco-textiles equals to fabrics made from cotton, silk or wool. Cotton as the largest natural fibre in the global fibre, textile and apparel economy represents now roughly 80 per cent of all natural fibres consumed. The natural fibres complex generates hundreds of billions of dollars of revenue annually in the global economy, and results in hundreds of millions of jobs. But it is important to point out, that regular cotton plants need a large land-use, high quantities of pesticides and fertilizers, and a high water consumption and there are also some ethical issues regarding genetically modified types.

Even regular wool has a high climate impact, and involves doubtful animal welfare practices (mulesing Merino sheep). Results of Life Cycle Assessments of environmental impacts of fibre production show that nowadays there are much more fibres on the market, which could be used for eco-fashion, than natural or organics. Fibres based on renewable primary products using eco-friendly production processes, or ones made of various forms of plant biomass or recycled material should not be ignored by designers for eco-fashion.

Eco-labels

A green product may be environmentally friendly in itself or produced and/ or packaged in an environmentally friendly way. Green labels could be important tools for decision making when purchasing textiles. Over the past ten years, the development of various eco-labelling schemes has raised the profile of environmental issues within the textile and clothing supply chain. Several national and international eco-labelling schemes are in use worldwide causing increasing confusion on the marketplace. Selected eco-labels for textiles were discussed by Author in another paper

INNOVATION – AS A KEY DRIVER FOR BEING SUSTAINABLE

The ever changing market forces the industry to respond very quickly. Consequently processes and technology change from one season to the other. New investments can be made only if they are in the focus of further development. Innovations may include new production technologies to reduce the labour requirement of garment completion and the development of novel 'smart' functions.

The most favoured innovation focuses on product development (functional and smart products) and on environmental friendly, cost effective process technologies. Innovation on the field of raw materials (nanofibres, functional fibres), knitting, weaving and nonwoven technology (space fabrics, tissue engineering), in finishing (digital printing, coating, nanotechnology, plasma treatment, etc.) are key drivers for the future of this branch. Producing innovative textiles could be a guarantee for success.

Chances for sustainable development of textile and clothing industry

A lot of efforts have been made worldwide to develop new environmental friendly processes in the textile chain. Experts in the UK recommend that for products in which raw material production dominates, in addition to measures to extend product life, alternative processes or materials should be pursued. A switch from conventional to organic cotton growing would eliminate most toxic releases, at the cost of price rises.

For products in which production dominates impacts, process efficiency should be pursued and the impact will be reduced by extending the life of the product or by re-using materials by some form of recycling. Sustainable alternatives:

- Use of organic fibres, recycled and eco-fibres based on eco-friendly production processes, etc.
- Use of new polymers: biomaterials/biopolymers, (reactive) hotmelts, laminates (nano)additives
- New finishes and coatings (metallized materials, super hydrophobicity, bioactive finishes, enzymatic processes, (atmospheric) treatments using cold plasma, liquid CO2, EB and UV, ultrasonic techniques, electrochemical dyeing)
- Lowering the environmental damages of wet processes – through modification, optimisation reducing consumption of dyestuff, chemicals, water and energy
- Process optimisation (Are all this process steps and auxiliaries in these quantities necessary?)
- Fast multi-step treatments
- Advanced textile wastewater treatment (use of algae as the bioremediation agent, heterogeneous complex, radiation, etc.)
- Reducing the amount of toxins and environmentally harmful contents in chemicals for dyeing, printing and finishing (alternative chemical products)
- Optimizing processes for laundries of hospitals and for dry-cleaning
- Saving energy in mechanical processes in apparel industry (energyefficient lighting)
- New technologies for the modification of the surface (plasma treatment)
- Implementing the sustainable approach in product development, in service and education.

The main effects of surface treatment using plasma is cleaning, the increase of microroughness (anti-pilling finishing of wool) and the hydrophilic surfaces. Fields of application are in desizing, functionalizing, and design of surface properties of textiles. Plasma treated textiles could be extremely dust- and dirtrepellent and hence should also be repellent to bacteria and fungi. It is successfully used for shrink-resist treatment of wool with a simultaneously positive effect on dyeing and printing.

Innovation such as self cleaning fabrics with super hydrophobe surface (lotus effect), the active climate control in cloth using Phase Change Materials

(PCM), shape memory alloys or polymers for thermal insulation, the functional fabrics protecting against heat, firer, radiation or mechanical stress and smart textiles with incorporated microelectronic devises for application in health care, emergency or sport open new perspectives and niche markets. Technology innovations such as 3D knitting and weaving may lead to economically viable production, with some consumer benefits from increased responsiveness.

However, this will only have environmental benefits if associated with material recycling. Comparisons in the energy consumption of clothes in the production and the use phase have shown that the use phase of clothing (washing, ironing, drying, etc.) needs by far the highest amount of energy, and causes the largest portion of CO2 emissions. Using innovative production techniques, wash temperatures of cotton products can be reduced and tumble drying avoided.

Novel treatments may provide resistance to odours so reducing the total number of washes or allow faster drying with less ironing. Technological innovation may lead to new means to freshen clothes without washing, efficient sorting of used clothing, new fibre recycling technologies and new low temperature detergents.

Research and developments in the Hungarian textile and clothing sector

Research activities related to the sustainability of the textile and clothing industry are being carried out at different knowledge centres in Hungary: at the Institute of Isotopes and at the Institute of Materials and Environmental Chemistry of the Hungarian Academy of Science, at BAY-ATI, Institute for Materials Science and Technology, and BAY-NANO, Institute for Nanotechnology, at the Department of Polymer Engineering and at the Faculty of Chemical and Bioengineering of Budapest University of Technology and Economics, at Faculty of Light Industry and Environmental Protection Engineering of Obuda University, further at INNOVATEXT Textile Engineering and Testing Institute Co., and at Nanovo Ltd., etc.

Two textile clusters were built in the past years in different regions of Hungary, the Pannon Textile Cluster (PanTex) and the South Great Plain Regional Cluster with the aim to help ensure the more efficient and profitable work of textile industry and service development specialists and enterprises. PanTex was established in 2005, with the aim to help ensure the more efficient and profitable work of textile industry and service development specialists and enterprises in the West Transdanubian Region.

The cluster offers his more than 37 members services in market research, trend tracking, career guidance, information service, interest representation, formation of a common website, project generation, benchmarking club, support for innovation, development activities, assistance in structuring supplier partner relations, partner meetings, assistance in participation at exhibitions and fairs, organization of business meetings, publishing of joint professional publications, arranging logistics activities, etc.

The South Great Plain Regional Textile Industry Cluster, founded in 2001 focusing on the knitting industry, have established their own supplier network, jointly operate a machinery service, coordinate orders in order to distribute capacity, operate a pattern design studio and put forward joint offers of products manufactured by the member enterprises. The cluster places great emphasis on product development. Their latest development is, for example, the application of nanotechnology in order to give their products an antibacterial element.

Besides the above mentioned in this chapter selected examples are presented for developments of last year realised in the Hungarian textile and clothing industry.

National Technology Platform for the Renewal of the Hungarian Textile and Clothing Industry (TEXPLAT)

The project was sponsored by the National Office for Research and Technology and coordinated by the Hungarian Society of Textile Technology and Science (TMTE). The aim of the project was to renew the Hungarian textile and clothing industry through the transformation of their product structure, as well as to determine the main directions of R and D and I activities accordingly for the next decade.

Besides uniting enterprises, TEXPLAT has a central role in the renewal of the product structure of the Hungarian textile and clothing industry, maintaining outside contacts of the industry as well as boosting the self-esteem and prestige of the sector. The platform focused on two milestone-like tasks in 2009-10 (in the first two years of its operation):

- to draw up a roadmap and a mission for the Hungarian textile and clothing industry as well as compiles its Strategic Research and Innovation Plan.
- to work out the Implementation Plan, based on the strategy.

An important task of the Platform is to involve the most competitive enterprises of the Hungarian textile and clothing industry – as many of them as possible, including SMEs too – in the activity of the Platform and to help them effectuate the results of their innovations in the form of marketable products and services.

Anti-bacterial Fabrics Using Nanotechnology

The Hungarian state-owned textile company, Szefo has jointly collaborated with Szeged University of Sciences to develop an anti-bacterial fabric by incorporating nano silver technology. The patented process and the anti-bacterial outer-wear fabric will be utilized in orders executed for Italy's Dama, its main exporting partner. Joining the Enterprise Europe Network, the 51-

employee small company, Eurotex Ltd, making knitwear for Italian designer Emidio Tucci among others, adapted also nanotechnology based finishing to produce jumpers lined with Russian-made nanosilver. Nanosilver has antibacterial substances designed to keep clothes fresher for longer. Since nanosilver-reinforced clothes prevent the wearer from sweating into the material, they require less frequent washing than other alternatives.

New Plant for the Production of Innovative Nonwoven Products

The automotive industry has become one of the most important industrial sector for Hungarian textile companies. The well-known German company, J.H. Ziegler GmbH is investing nearly 7 million EUR in a new production building and in the most modern, largest and most versatile production systems for nonwovens in Bábolna. The company opened his first plant in Hungary in 2006. It has been producing nonwovens primarily for the furniture industry, but it will now be able to supply the automotive sector with a far greater proportion of acoustic absorbers and other nonwoven components directly from Bábolna/ Hungary. In 2010, Ziegler achieved a turnover of around 30 million EUR, which represents a 40 per cent increase compared to 2009. Just over half of the total turnover came from the automotive industry and its sub suppliers. Background of the investments are:

- Good productivity of the workforce,
- Excellent logistics location (close to the city of Gyõr)
- Continued growth of the local automotive industry

Increase of capacities to produce innovative active wear

Manifattura Italiana Tessuti Indemagliabili Spa (M.I.T.I.), an Italy-based manufacturer of high-quality fabrics, is setting up a plant in Hungary with innovative production facilities. The company has refurbished a textile plant in Szentgotthárd (western Hungary), where it proposes to increase its concentrated staff strength from 40 to 100 employees by adding other 60 employees. It invested over 2 million for the refurbishment of the 4,000-square-metre production capacity to furnish it with high-tech weaving machinery. Cheap availability of labour in Hungary was the primary reason for this shift.

M.I.T.I. is to set on board in Hungary, a new development scheme which apart from enhancing the capacity will also expand its range of products. Over the past 78 years, M.I.T.I. has produced several different kinds of knitted fabrics. Also during the past decades its activities have concentrated on high-performance stretch fabrics for active wear. With its 200 employees its main focus is on the western European market.

THE CHALLENGE OF TEXTIL AND DESIGN EDUCATION AT RKK

Education plays an important role in forming designer's approach sustainable. Its task is to explain the dual challenge of designers - to improve the level of functions, to constantly recreate new products while at the same time to save the environment and to avoid disturbing it. An important task of our textile design education is to show the basics of eco-design - how principles of ecological, ethical and social aspects could be combined during the design process. As a teacher of textile and marketing and as researcher I would like to contribute to the changes of our design education highlighting the importance of our responsibility for the future. Students need professional clothing design skills, but also adequate information on the actual problems of sustainability in the textile and fashion industry.

It is important to explain them the disadvantages of global business - it causes tremendous environmental problems due to the long transit ways, etc. -, and to point out the social and ethical responsibility (the trend of over-consumption of clothes creating "fashion victims"), the poor working conditions in sweatshops and, among others, the principles of fair- trade. Increased emphasis on durability as a component of fashion would support a move towards reduced material flow. The sector could have its material flow without economic loss if consumers pay a higher price for a product that lasts twice as long. Our education has encouraged several activities in the last years related to the topic sustainability and eco-design. A successful initiative for advertising ecodesign was the competition for schools organised at RKK in the past years. Interesting design concepts for recycling and reuse were sent from 11 Hungarian high schools to the exhibition in 2010 at RKK.

Fig. "Waste couture" by Dobos Emese, Green design Competiti on for high school students at RKK

The protection of the environment was one of the most important topics at the summer university programmes "European Digital Print Media" . The international project was organised three times at "Rejto Sandor" Faculty of Light Industry and Environmental Engineering (RKK) at Obuda University with the financial support of the European LLP/Erasmus programme. Innovative digital printing technologies were used to form designer approach responsible for the future. At the final exhibition of the 3-week programmes students presented the importance of sustainability in different ways.

Fig. Student's works using digital printing techniques

2

The Role of Service-Learning and Training

DEFINITION

The term training refers to the acquisition of knowledge, skills, and competencies as a result of the teaching of vocational or practical skills and knowledge that relate to specific useful competencies. It forms the core of apprenticeships and provides the backbone of content at institutes of technology.

In addition to the basic training required for a trade, occupation or profession, observers of the labour-market recognize today the need to continue training beyond initial qualifications: to maintain, upgrade and update skills throughout working life. People within many professions and occupations may refer to this sort of training as professional development. Some commentators use a similar term for workplace learning to improve performance: training and development.

One can generally categorize such training as on-the-job or off-the-job:

- On-the-job training takes place in a normal working situation, using the actual tools, equipment, documents or materials that trainees will use when fully trained. On-the-job training has a general reputation as most effective for vocational work.
- Off-the-job training takes place away from normal work situations—implying that the employee does not count as a directly productive worker while such training takes place. Off-the-job training has the advantage that it allows people to get away from work and concentrate more thoroughly on the training itself. This type of training has proven more effective in inculcating concepts and ideas.

Training differs from exercise in that people may dabble in exercise as an occasional activity for fun. Training has specific goals of improving one's capability, capacity, and performance.

SUPERVISORY TRAINING METHODS

Training new supervisors involves preparing newly promoted managers to assume a leadership role. Methods for ensuring that supervisors develop the correct skills include providing lectures, self-paced materials and coaching or mentoring opportunities. In addition to training new supervisors, training programmes typically include refresher courses as well as fundamental skills curricula. Supervisory training courses involve practicing interpersonal skills and communicating effectively, both verbally and in written form.

PLANNING

Developing effective supervisory training involves defining an overall organizational plan. Ideally, your plan reflects current organizational needs and links to operational metrics. List eight to 10 learning objectives for the programme such as being able to complete performance reviews, control department costs or handle insubordination.

Including a self-assessment allows supervisors to determine their own learning needs, relative to becoming an effective supervisor in your organization. Working with former peers can be a daunting task and new supervisors tend to need extensive support as they transition to their new company role.

LEARNING ACTIVITIES

Effective supervisor training begins with an overview of management concepts such as planning, organizing, motivating, controlling and scheduling. Providing ample opportunity to role-play difficult situations enhances a new supervisor's capacity to respond well in the work environment. Learning to use management reports requires expertise with computer systems and their related procedures, so providing practice lab environments creates the best way to develop these skills.

Some supervisory tasks involve complying with local and federal regulations, such as interviewing, hiring and terminating employees. These tasks typically practice studying sample cases to develop skills in recognizing how to handle difficult and complex situations associated with functioning as a supervisor.

ONGOING SUPPORT

Providing ongoing for support for supervisors ensures your managers can continue to develop their skills. By scheduling regular lectures or workshops, you can allow supervisors to build on existing skills. Encouraging supervisors to seek out peer advice also builds an effective community.

ON-THE-JOB METHODS

APPRENTICESHIP PROGRAMME

An apprenticeship programme is a course of education that is based on on-the-job experience. Furthermore, an apprenticeship programme usually focuses on one trade or skill rather than on a variety of subject areas. For example, someone trying to get started in event photography might apprentice for a photographer who makes a living shooting events. Alternatively, someone interested in learning how to become a car mechanic might apprentice at a garage.

Apprenticeship training can take anywhere from one to six years, depending on the trade. In general, you have to be eighteen years old to begin an apprenticeship programme. However, it is possible to begin apprenticing at the age of sixteen with parental permission. Apprentices function as both assistants and shadows to their employers. Because of the nature of apprenticeships, it is possible to learn a great deal about an industry in a rather short amount of time. Furthermore, all of the learning is hands-on rather than theoretical. While apprentices may have to take care of a good deal of menial tasks, they also get the ability to watch professionals in action, which is a great way to learn a trade. One unique aspect of apprenticeship programmes is that, while learning a particular trade, it is also possible to learn the business behind the trade.

One of the major benefits of participating in an apprenticeship programme is that you can earn money while you are learning a trade. It is important to note, however, that there are many employers in the market who, although they might be impressed with the credentials that you will gain in an apprenticeship programme, require a diploma from an accredited institution in order to consider you for a position. Research the field that interests you and find out what employers are looking for. If most employers are looking for a two year degree or a four year degree, consider including an apprenticeship programme in your degree. It is possible that you will be able to earn college credit for your work as an apprentice.

Many colleges offer credit to students who complete apprenticeship and internship programmes. Many states offer information about apprenticeship programmes on their Department of Labour and Industries websites. Consider researching the information on your state's website or calling your Department of Labour. The United States Department of Labour website also has information that is helpful to people throughout the country about finding jobs and participating in apprenticeship programmes.

JOB INSTRUCTION TRAINING

Job Instruction Training is a logical outgrowth of Job Hazard Analysis. It is a proven technique for teaching new skills and safe, healthful work habits faster and more effectively. All new employees and those transferred to new jobs should receive JIT. One of the first steps is trainer selection—preferably a supervisor or a skilled person within the department.

Regardless of who is selected, the trainer should:

- Know the job in question thoroughly
- Have leadership skills

- Have a desire to teach others
- Be friendly and cooperative
- Have a professional attitude towards the job and other employees

Prepare to Instruct

- *Have A Time Table*: How much skill should the trainee have by what date.
- *Break Down The Job*: List important steps and key points (safety is always a key point). Use a Job Hazard Analysis (JHA) breakdown to locate and identify hazards.
- *Have Everything Ready*: The right equipment, materials and supplies should be in place and ready to go.
- Have The Workplace Properly Arranged, just as the trainee will be expected to keep it.

Write the steps of each job in sequence, noting the safest, most efficient way. A thoroughly prepared Job Hazard Analysis will provide much of this. Estimate how long a JIT session will last. Allow enough time in your schedule, depending on the complexity of the job and the trainee's previous knowledge and experience.

How to Instruct

Job Instruction Training should proceed in four steps (often referred to as the Four-Point Method.)

Prepare the Worker

- Put the trainee at ease
- Define the job and find out what is already known about it
- Get the employee interested in learning the job
- Place yourself in the correct position

Since this is a one-on-one experience, start by putting the worker at ease. Explain all responsibilities and procedures. Show the employee how the job contributes to the overall work of your firm—how it fits in. Emphasize the need for quality, production and safety.

Present the Operation

- Tell, show and emphasize ONE IMPORTANT STEP at a time
- Stress each KEY POINT (Safety is ALWAYS a key point)

Position yourself alongside the trainee so that he or she will see the job as it is done and not in reverse.

Demonstrate and explain as you're doing the job. Ask the trainee to explain the process to you. If something has been missed or misunderstood, go back over it at once. Demonstrate the use of all required personal protection equipment and tell why machine guards are important. Explain thoroughly all personal safety regulations. Encourage the employee to ask questions. Training should reflect intelligent conversation, not a sermon.

Try Out Performance

- Have the employee do the job—provide coaching and correct any errors
- Have the employee explain each KEY POINT to you during the process
- Make sure the worker understands
- Continue until YOU know the worker fully understands

Once you're sure the trainee understands the operation, it's time for a tryout under your careful supervision. As he or she performs the operation, have the worker explain each step, including the reasons "why" things were done. If the employee makes mistakes, explain calmly how to do things right. Work patiently with the employee until each step is mastered.

Follow-up

- Let the employee work independently
- Designate a person to go to for assistance
- Check frequently, encourage questions
- Taper off extra coaching and close follow-up

Job Instruction Training is a continuing process. Follow-up from time to time to be sure things are going well. Be sure the new employee knows where to find help.

OFF-THE-JOB METHODS

CLASSROOM LECTURE TRAINING

Requirements

- Realise that to get the most out of CLT, you must put something into it. For this reason, we have seen the need for all staff who participate in CLT to make the following four training commitments. These commitments are listed on the Registration/Commitment form.
  - Prepare a 30 to 35 minute lecture before the CLT week.
  - Give the lecture in a class during the CLT week.
  - Remain on location and be involved during the entire week (usually Sunday afternoon to 5 p.m. Friday).
  - Follow through by speaking in at least one class per quarter for the remainder of the year.
- *Note:* If your whole team will be involved in CLT, your director has probably already made these commitments for you. If your director has not, you will need to make them yourself. Return this completed

Registration/Commitment form to your local CLT coordinator before the registration deadline.
- Remember that these requirements are designed to help you and your ministry receive maximum benefit.

Available Materials

Familiarize yourself with the following resources.
- A set of two tapes.
  - The first side of tape one is an actual classroom lecture. The rest of the material on the tapes covers how to prepare and present a lecture. Each campus involved in classroom training should receive one set of tapes from the local CLT coordinator. They should be listened to both before and after preparation of your talks.
  - *This training manual, Which includes the following:*
    (a) Staff Orientation Sheet.
    (b) Registration/Commitment Form.
    (c) How to Prepare and Present Classroom Lectures.
    (d) How to Line Up Classroom Meetings.
    (e) How to Write a Topic List.
    (f) Collegiate Challenge reprint on the 10 most frequently asked questions.
    (g) How to Follow up Classroom Lectures.
    (h) How to Involve Others in Classroom Lecturing.
    (i) Classroom Lecture Critique Sheet.
    (j) Speaker's Self-evaluation Form.
    (k) Action Steps.
    (l) Order Forms for tapes, classroom lecture outline manual and related materials.
- Classroom lecture tapes.
  - A series of 18 tapes on various academic subjects.
- University Classroom Lecture
  - A collection of detailed classroom lecture outlines complete with documentation, recommended reading and questions most asked on each subject. An substance by Chris Hall on the legality of Christian activity on secular campuses and letters of recommendation to present to professors are also included.
- Refer to the materials frequently for guidelines and topic ideas.

Preparation for CLT

- Pray
  - As you pray, anticipate results. "The effective prayer of a righteous man can accomplish much".

- Consider these suggestions for prayer.

- Pray as a staff team, with students, etc.
- Put CLT prayer requests in your personal newsletters. Print answers after CLT is over.
- Contact local churches, prayer chains, etc. for prayer.
- Pray for boldness, efficiency in preparation and open-mindedness of students and professors. Pray that the staff on the host campus will be able to line up one class for each participant. Pray for Satan to be bound and God to be glorified.
- Listen to training tapes.
  - Note that the tapes are designed to follow and supplement the handouts in this manual. Listen to the tapes before and after preparation of your lecture. Each participant should have listened to the training tapes at least twice before CLT week.
- Select a topic.
- Gather information.
- Prepare lecture
  - Consider using University Classroom Lecturing for outline ideas. These outlines are well-documented and include recommended reading and the questions most often asked on each subject.
  - Plan your lectures to be 30-35 minutes long.
  - See the Classroom Lecture Critique Sheet for pointers on evaluation.
  - Complete your talk before CLT week begins. This is a must. There is no time scheduled during the week for preparation.
- Practice Lecture.
  - Consider practicing your lecture on some of your staff or students before the CLT week. The Speaker's Self-evaluation sheet presents a simple format for a practice evaluation.
- Copy lecture.
  - Photocopy your completed lecture and turn it in at the first CLT meeting.
  - Additional Notes – Questions You May Have
  - Should I being this orientation notebook to CLT?
    (a) Yes. You will use the materials and can use the notebook for any additional handouts you receive.
  - What will the schedule be like?
    (a) We have an "ideal" schedule, but the actual schedule must be determined by the times of the classes in which we are able to speak. An overview of the ideal schedule is:
    (b) Sunday afternoon or evening Orientation

(c) Monday Seminars
(d) Tuesday Practice lectures; observe trainer speak in classes
(e) Wednesday Observe trainer speak (possibly you speak)
(f) Thursday and Friday You speak in classes and are evaluated by staff
(g) A few other seminars are included on Tuesday through Friday. We do our best to schedule breaks and some recreation.
(h) Be sure to come rested and prepared!

– If I am on the host campus, do I continue my normal ministry activities during CLT week?
  (a) You will need to flex. Since the week is usually full (Sunday afternoon to Friday afternoon) and since the schedule is sometimes not finalized until the first day of CLT, we advise against keeping a full schedule of regular appointments, etc. Talk it over with your campus director. We suggest keeping action group meetings, LTC's, College Life meetings, etc., as long as they do not conflict with CLT. You might bring students with whom you meet regularly to the CLT function scheduled at that time or to your class to hear you speak.
– If I am coming in from out of town, Where will I stay?
  (a) You should make housing arrangements through the local CLT coordinator. Staff from other campuses should plan to stay on location. Even commuting one to two hours one way every day tends to wear you out. If you have preferences for motels versus private homes (or *vice versa*), be sure to voice them to the coordinator. However, please realise that for various logistical or PR reasons, the coordinator sometimes may need to assign everyone to a particular housing arrangement.
– Do I have to pay a registration fee?
  (a) There is a basic fee of $30 per staff (to cover materials, tapes, administrative costs, etc.). Some campuses, in order to share expenses, work out an agreement for each staff member or campus to pay an additional amount. Check with the local coordinator. Fees will be collected at the CLT.
– Will I enjoy the CLT week?
  (a) You will enjoy CLT week!

SIMULATION EXERCISE

- An exercise is a practice activity that places participants in a simulated situation requiring them to function in the capacity expected of them in a real event.

- Its purpose is to promote preparedness by testing policies and plans, SOPs, and personnel training.

Why Conduct a Simulation Exercise

- Simulation exercises are conducted to evaluate an organization's ability to execute one of more portions of it's response.
- Many successful responses to emergencies are attributed to previous simulation exercises.

Commonalities of Disaster after Action Reports

- Planning, training and exercising are the only feasible recommendations.
- Relationships must be established, plans written and tested, and procedures agreed upon.
- Effective coordination cannot be achieved during the chaos of an actual event.

Main Benefits of Exercise Simulations

- *Individual training*: Allow people to practice their roles, gain experience in their roles without an actual disaster.
- *System Improvement*: Improves the organizations system for managing emergencies.
- *Benefits*: Occur from actual practice and evaluations, and then following through with recommendations.

Reasons to Conduct Exercise Simulations

- Test and evaluate plans
- Reveal planning weaknesses
- Reveal gaps in resources
- Improve organizational coordination
- Clarify roles and responsibilities
- Improve individual performance
- Gain recognition and support of officials
- Satisfy regulatory requirements
- Train personnel

Five Main Types of Simulation Exercises

- Orientation Seminar
- Drill
- Tabletop Exercise
- Functional Exercise
- Full-Scale Exercise

Orientation Seminar

- An overview or introduction.
- Purpose is to familiarize participants with roles, plans, procedures or equipment.
- Can be used to resolve questions of coordination and assignment of responsibilities.

Drills

- A coordinated, supervised exercise activity, normally used to test a single specific operation or function.
- There is no attempt to coordinate organizations.
- Practice and perfect one small part of the response plan.
- The effectiveness is its focus on a single, relatively limited portion of an overall system

Tabletop Exercise

- Facilitated analysis of an emergency situation in an informal stress-free environment.
- Designed to elicit constructive discussion as participants examine and resolve problems based on existing operational plans and identify where plans need to be refined.
- Success is determined by group participation and problem identification.
- Minimal attempt at simulation: equipment is not used, resources are not deployed, and no time pressures are introduced

Functional Exercise

- A fully simulated interactive exercise that tests the capability of an organization to respond to an event.
- Tests multiple functions of the organization's operational plan.
- A coordinated response to a situation in a time pressured, realistic simulation.
- Focuses on the coordination, integration, and interaction of an organization's policies, procedures, roles and responsibilities before, during, or after the simulated event.

Full-Scale Exercise

- Simulates a real event as closely as possible.
- Designed to evaluate the operational capability of emergency response.
- Conducted in a stressful environment that simulates actual response conditions.
- Requires the mobilization and actual movement of emergency personnel, equipment, and resources.

Simulation Exercise Design

Simulation exercises model a common workplace scenario and allow for problem solving. A common simulation exercise is the in-box exercise where participants sit at simulated desks and are given an in-box of documents common to their role. They are then instructed to determine their actions using this information and may be given road-blocks along the way to again simulate the real work environment where distractions and changing priorities are a fact of life. Simulation training is often used for emergency preparedness training.

Here an emergency is simulated and everyone from volunteers, to first aid attendants, to administrators can practice procedures to ensure they are understood and are realistic. It is a great way to be sure all the bases have been covered.

Simulation training is commonly used in retail training. Some companies go to the length of creating a dummy store that simulates a real store in every way except that it is not open to the public. This allows trainers to teach all aspects of the business without compromising customer service or inconveniencing staff at a fully functioning outlet. The other use of a dummy store is to test out new systems or procedures to make sure they work before introducing the change to the real stores.

Actors are sometimes hired to role play in simulation exercises. The great thing about hiring an actor is that they are unknown to the staff and can act out distractions and twists and turns convincingly. It is a great way to give a student actor some extra cash!

CASE PRESENTATION

A case presentation is a formal communication between health care professionals (doctors, pharmacists, nurses, therapists, nutritionist etc.) regarding a patient's clinical information.

Essential parts of a case presentation include:

- Identification
- Reason for consultation/admission
- Chief complaints (CC)—what made patients to seek medical attention
- History of present illness (HPI)—circumstances relating to chief complaints
- Past medical history (PMHx)
- Past surgical history
- Current medications
- Allergies
- Family history (FHx)
- Social history (SocHx)
- Physical examination (PE)
- Laboratory results (Lab)

- Other investigations (imaging, biopsy etc.)
- Case summary and impression
- Management plans
- Follow up in clinic or hospital
- Adherence of the patient to treatment
- Success of the treatment or failure
- Causes of success or failure.

EXPERIENTIAL EXERCISES FOR TEACHING STRATEGIC MANAGEMENT

THE BLUE CHIP GAME

Teaching objective: To introduce the concepts of game theory, cooperation, collusion, opportunism, and to reinforce learning about framing and boundaries.

Time: approximately 15 minutes

Materials: 8 index cards (4 blue, and 4 of another colour—*e.g.*, yellow ones), and 4 envelopes (preferably thick enough to conceal the content of the cards). If you prefer, you can use poker chips.

Overview: Each team submits one card each round; and earns/(loses) points depending on what the other teams do. If everyone submits a yellow card, then everyone wins. But if only one group "defects", then it does best of all. But if everyone defects, everyone loses points.

Prior to class: There are four teams in class: each team gets an envelope, and one card of each colour. I number the cards and envelopes, so I can keep track of the game more easily. You'll also want to make up a viewgraph, that shows how the game is scored.

Cards turned in	Blue is worth	Yellow is worth
0 Blue and 4 Yellow	0	25
1 Blue and 3 Yellow	75	– 25
2 Blue and 2 Yellow	50	–50
3 Blue and 1 Yellow	25	–75
4 Blue and 0 Yellow	–75	0

In Class: Divide class into four teams, have people set with their groups. Announce the following.

"This game has four rounds. In each round you'll have 60 seconds to turn in one card, either a blue card, or a yellow card. Your score is determined partially by what you choose, and partially by what the other teams choose. Your objective is to score as many points as possible.

There is one rule: You must turn in, using the envelopes provided, 1 card each round. (Note: this rule is optional.) Turning in two or turning in zero will cost you 100 points. Good luck."

After each round, I post the current scores, as well as the cumulative scores for each team. I like to remind students of the rules and the objective, and will banter jokes with teams. After each round I announce there is a secret for winning. Roughly half the time the groups will figure out they can collude with other groups. For the times when they don't, I ask students what the rules are; after some prodding, they acknowledge that collusion is not illegal in this game. About half of the time the groups collude successfully, and half the time they don't.

Sample discussion questions:

- How do you win at this game?
- At what point did you realise you could cooperate with other teams?
- What rules did you assume existed that didn't?
- For the "industry" which scenario is most lucrative? How about for individual teams?
- What steps could you have taken to ensure cooperation occurred?
- What made cooperation more difficult?
- Are any teams "untrustworthy", or are they rational?
- How might you address "defections".
- What were the more interesting strategies?
- Can you think of any situations where this situation exists (*e.g.* cartels like OPEC).
- Can cartels succeed?
- How would you play the game differently now? (Sometimes I repeat this game a week or so later).

When I announce a "trick exists to win", I get some really creative efforts. Here are 2 interesting ones:

1. I had a football player in class, polite, but intimidating. The teams wouldn't cooperate and he got frustrated. In the 4th round he went to the other groups, and told them to submit the yellow card. Each group, feeling coerced, did so. Later, they complained of being pressured. I gave them one of my favourite quips; "Coercion is just cooperation by other means" (okay, I guess you had to be there).
2. I had a group cut its cards in half, and taped the mismatched pairs together—that way they could claim it was yellow (provided everyone else was yellow), or blue (in case it was a split decision)

PUTTING TOGETHER THE STRATEGY PUZZLE

Putting Together The Strategy Puzzle objectives:

- The importance of being explicit about and challenging existing operating assumptions.
- The importance of understanding what resources are available
- The importance of recognizing and utilizing commonly overlooked resources/expertise within the firm.
- The value of big picture vs. in the trenches perspectives.

Set-up

Requires two 25-piece puzzles. Prior to the exercise, two pieces from each puzzle should be blacked-out with an indelible marker. Additionally, one of the fifty pieces should be placed in the trainer's pocket prior to the training. Note: I like using the puzzles that have three concentric circles depicting people or animals. The inner circle is the face, the mid part is the body, and the outer circle is the legs and feet. The circles can be rotated to place a dog's head on an elephant's body etc. once the puzzle is constructed. When participants enter, there should be a few pieces at each of their places.

Group Instructions

Instructor says:

- "Your task is to put all of the puzzle pieces together using all of the resources available to you in the room. You will have 3 minutes. OK, go."
- *What happens*: The group eventually figures out there are two puzzles. They often leave the black pieces aside as "not belonging to the puzzle" or figure out their place in the puzzles at the last moment after initially declaring "we're done." The also declare one piece to be missing.

Discussion Questions:

- What was this experience like for you?
- What did you assume initially about the number of puzzles? What made you think the black pieces did not belong? Remind them that the ir task was to put all of the puzzle pieces together"
- What assumptions did you make about what the puzzle should look like? Why?
- What resources did you use?
- What resources did you not use? Who in the room did you not ask for input?
- Some of you worked hands-on during the task of puzzle building. What was this like? What was unique about your perspective?
- Others stood back and gave directions from a distance. What was this like? What was unique about your perspective?
- (ask only if your have circular puzzle) If we rotate the parts of the puzzles do the pieces still fit? Yes... So why did you not construct the puzzle to show a dog's head on an elephants body? It certainly would have met the goal I set for you...

Key Lecture Points

- Preconceived ideas hurt process. We often have preconceived ideas that may in fact, be irrelevant to the true business goal. Similarly, putting together all of the pieces is more important than your preconceive notion of aesthetics. It's important to get these mental maps on the table during planning.

- Ignoring key expertise. Often people in organizations who can make a contribution of expertise or experience get overlooked because they have not traditionally played a role in strategy development. They may hold the missing piece to a winning strategy.
- Time horizons. Inevitably, people on the front lines (*e.g.*, operations) have a very different view than the long- view people (*e.g.*, managers and strategists). Firms need both perspectives in strategic management.
- More than one way. There is often more than one way to achieve a goal. Firms get caught up in picking the perfect strategy when, less elegant strategies are quite effective (*e.g.*, dog head on elephant's body).

The Paper Chase

The paper fight is an exercise that simulates hyper-competition. Strategies emerge but are perfectly imitable. As a result, a strategy will not create an advantage beyond one period.

Each round might also be used to simulate different stages of the industry life cycle.

Preparation:

- Place two uneven piles of paper on each side of the room (uneven allows you to talk about different resource endowments).
- Invite two groups to the front (on each side)
- With no warning or explanation, say, "We are going to have a paper fight. Ready, go!"
- Let them go for a short period of time (until you see a strategy of any kind)
- Have a discussion about what happened.
- Invite two other groups (or another group to challenge the winner) to do it again. Possibly repeat a third time.

What happens the first round: The first time, the groups will hesitate. There are no rules—this is an undefined landscape (embrionic industry). Sometimes, it will not even be clear that they should "organize" as two "firms" and they will fight amongst themselves.

Generally, they will attack the other team after a short hesitation. However, they will often hold their position more than chase the other team around the room (no flanking maneuvers, or more complex strategies). The organization will either have the same people crumpling and throwing or people will specialize in crumpling or throwing.

Discussion questions:

- Why did you fight? Why did you hesitate?
- Who won? Why do you say so? (criteria for success/org performance—No. of paper wads on the other side *vs.* No. of hits scored)
- What were the internal structures of the "firms"?
- Resources—were the teams on equal footing? What resources mattered (not the amount of paper)?
- What strategies emerged? Did it matter? How did the other side respond?

What happens the second round: The second time we tend to see more complex strategies and organizations. The landscape is changed and they can now throw back the crumpled wads from the last bout. Strategies include: (1) have everyone throw back wads and no one crumples, (2) flanking maneuvers, (3) steal the other team's paper, (4) movement around the room.

Discussion questions:

- What was different this time? How did the rules change?
- What does it take to win now? Would the same strategy work again?
- What were the internal structures of the "firms"?
- What would happen if we ran it again? And after that? How long is a strategy useful? (note for each strategy, there is a response that neutralizes it)

COMPUTER MODELING

This is one of the most powerful tools available to science and engineering and, like all powerful tools, it brings dangers as well as benefits. Andrew Donald Booth said, "Every system is its own best analogue." As a scientist he should, of course, have said in what sense he means best. The statement is true in terms of accuracy but not in terms of utility. If you want to determine the optimum shape for the members of a bridge structure, for example, you cannot build half a dozen bridges and test them to destruction, but you can try large numbers of variations in a computer model.

Computers allow us to optimise designs in ways that were unavailable in times past.

Nevertheless, the very flexibility of a computer programme, the ease with which a glib algorithm can be implemented with a few lines of code and the difficulty of fully understanding its implications can pave the path to Cloud Cuckoo Land.

ASSUMPTIONS

At almost every stage in the development of a model it is necessary to make assumptions, perhaps hundreds of them. These might or might not be considered reasonable by others in the field, but they rapidly become hidden.

Some of the larger models of recent times deal with the interactions of variables whose very nature is virtually unknown to science.

AUDITABILITY

In olden times, if a scientist published a theory, all the stages of reasoning that led to it could be critically examined by other scientists. With a computer model, it is possible, within a few days of development, for it to become so complex that it is a virtual impossibility for an outsider to understand it fully. Indeed, where it is the result of a team effort, it becomes unlikely that any individual understands it.

OMISSIONS

Often vital elements can be left out of a model and the effect of the omissions is only realised if and when it is tested against reality. A notorious example is the Millennium Bridge in London. It was only after it was built and people started to walk on it that the engineers realised that they had created a resonant structure. This could have been modelled dynamically if they had thought about it. Some models that produce profound political and economic consequences have never faced such a challenge.

SUBCONSCIOUS

The human subconscious is a powerful force. Even in relatively simple physical measurements it has been shown that the results can be affected by the desires and expectations of the experimenter. In a large computer model this effect can be multiplied a thousandfold. Naturally, we discount the possibility of deliberate fraud.

SOPHISTICATION

This word, which literally means falsification or adulteration, has come to mean advanced and efficient. In large computer models, however, the literal meaning is often more applicable. The structure simply becomes too large and complex for the inputs that support it.

TESTABILITY

When we were pioneering the applications of computer modelling about forty years ago, we soon came to the cease that a model is useless unless it can be tested against reality. f a model gives a reasonably accurate prediction on a simple system then we have reasonable, but not irrefutable, grounds for believing it to be accurate in other circumstances. Unfortunately, this is one of the truisms that have been lost in the enthusiasms of the new age.

CHAOS

Large models are often chaotic, which means that very small changes in the input variables produce very large changes in the output variables. Some very simple processes can amplify errors, taking the difference between numbers of a similar magnitude for example. The errors (or noise) are then propagated through the system. If there are feedback mechanisms present, it is quite possible for systems to operate on the noise alone. Many of the computer models that receive great media coverage and political endorsement fail under some of these headings; and, indeed, some fail under all of them. Yet they are used as the excuse for profound, and often extremely damaging, policies that affect everyone. That is why computer models are dangerous tools.

SIMULATION VERSUS MODELING

Traditionally, forming large models of systems has been via a mathematical model, which attempts to find analytical solutions to problems and thereby enable the prediction of the behaviour of the system from a set of parametres and initial conditions.

While computer simulations might use some algorithms from purely mathematical models, computers can combine simulations with reality or actual events, such as generating input responses, to simulate test subjects who are no longer present. Whereas the missing test subjects are being modeled/ simulated, the system they use could be the actual equipment, revealing performance limits or defects in long-term use by these simulated users. Note that the term computer simulation is broader than computer modeling, which implies that all aspects are being modeled in the computer representation. However, computer simulation also includes generating inputs from simulated users to run actual computer software or equipment, with only part of the system being modeled: an example would be flight simulators which can run machines as well as actual flight software. Computer simulations are used in many fields, including science, technology, entertainment, health care, and business planning and scheduling.

DATA PREPARATION

The data input/output for the simulation can be either through formatted textfiles or a pre- and postprocessor. Data preparation is possibly the most important aspect of computer simulation. Since the simulation is digital with the inherent necessity of rounding/truncation error, even small errors in the original data can accumulate into substantial error later in the simulation.

While all computer analysis is subject to the "GIGO" (garbage in, garbage out) restriction, this is especially true of digital simulation. Indeed, it was the observation of this inherent, cumulative error, for digital systems that is the origin of chaos theory.

TYPES

Computer models can be classified according to several independent pairs of attributes, including:

- Stochastic or deterministic (and as a special case of deterministic, chaotic)
- Steady-state or dynamic
- Continuous or discrete (and as an important special case of discrete, discrete event or DE models)
- Local or distributed.

Equations define the relationships between elements of the modeled system and attempt to find a state in which the system is in equilibrium. Such models are often used in simulating physical systems, as a simpler modeling case before dynamic simulation is attempted.

- Dynamic simulations model changes in a system in response to (usually changing) input signals.
- Stochastic models use random number generators to model chance or random events;
- A discrete event simulation (DES) manages events in time. Most computer, logic-test and fault-tree simulations are of this type. In this type of simulation, the simulator maintains a queue of events sorted by the simulated time they should occur. The simulator reads the queue and triggers new events as each event is processed. It is not important to execute the simulation in real time. It's often more important to be able to access the data produced by the simulation, to discover logic defects in the design, or the sequence of events.
- A continuous dynamic simulation performs numerical solution of differential-algebraic equations or differential equations (either partial or ordinary). Periodically, the simulation programme solves all the equations, and uses the numbers to change the state and output of the simulation. Applications include flight simulators, construction and management simulation games, chemical process modeling, and simulations of electrical circuits. Originally, these kinds of simulations were actually implemented on analog computers, where the differential equations could be represented directly by various electrical components such as op-amps. By the late 1980s, however, most "analog" simulations were run on conventional digital computers that emulate the behaviour of an analog computer.
- A special type of discrete simulation which does not rely on a model with an underlying equation, but can nonetheless be represented formally, is agent-based simulation. In agent-based simulation, the individual entities (such as molecules, cells, trees or consumers) in the model are represented directly (rather than by their density or concentration) and possess an internal state and set of behaviours or rules which determine how the agent's state is updated from one time-step to the next.

- Distributed models run on a network of interconnected computers, possibly through the Internet. Simulations dispersed across multiple host computers like this are often referred to as "distributed simulations". There are several standards for distributed simulation, including Aggregate Level Simulation Protocol (ALSP), Distributed Interactive Simulation (DIS), the High Level Architecture (simulation) (HLA) and the Test and Training Enabling Architecture (TENA).

VESTIBULE TRAINING

In the early 1800s, factory schools were created, due to the industrial revolution, in which workers were trained in classrooms within the factory walls. The apprentice system was inadequate due to the number of learners that had to be trained as the machines of the Industrial Revolution increased the ability of the factory to produce goods.

The factory owners needed trained workers quickly because there was a large demand for the produced goods. Towards the end of the 1800s, a method that combined the benefits of the classroom with the benefits of on-the-job training, called vestibule training, became a popular form of training.

The classroom was located as close as conditions allowed to the department for which the workers were being trained. It was furnished with the same machines as used in production. There were normally six to ten workers per trainer, who were skilled workers or supervisors from the company. There are many advantages of vestibule training. The workers are trained as if on the job, but it did not interfere with the more vital task of production. Transfer of skills and knowledge to the workplace was not required since the classroom was a model of the working environment. Classes were small so that the learners received immediate feedback and could ask questions more easily than in a large classroom. Its main disadvantage is that it is quite expensive as it duplicates the production line and has a small learner to trainer ratio.

PROGRAMMED LEARNING

Programmed Learning or Programmed Instruction is a learning methodology or technique first proposed by the behaviourist B. F. Skinner in 1958. According to Skinner, the purpose of programmed learning is to "manage human learning under controlled conditions".

Programmed learning has three elements:

1. It delivers information in small bites,
2. It is self-paced by the learner, and
3. It provides immediate feedback, both positive and negative, to the learner.

It was popular in the late 1960s and through the 1970s, but pedagogical interest was lost in the early 1980s as it was difficult to implement and its limitations were not well understood by practitioners. It was revived in the 1990s in the computerized Integrated Learning System (ILS) approach, primarily in the business and managerial context. The methodology involves self-administered and self-paced learning, in which the student is presented with information in small steps often referred to as "frames". Each frame contains a small segment of the information to be learned, and a question which the student must answer. After each frame the student uncovers, or is directed to, additional information based on an incorrect answer, or positive feedback for a correct answer.

EXAMPLES

Daily Oral Language and the Saxon method, a math programme, are specific implementations of programmed instruction which have an emphasis on repetition. Well-known books using programmed learning include the Lisp/ Scheme text, *The Little Schemer and Bobby Fischer Teaches Chess.*

CRITICISM

Programmed Instruction has been criticized for its inability to provide adequate feedback on incorrect answers and for its lack of student instigated conceptualization opportunities. It works best in basic courses which introduce the vocabulary of a discipline, heavily fact-based courses, and rule-based technical courses.

CONFERENCE

A conference is a meeting of people that "confer" about a topic:

- Academic conference, in science and academia, a formal event where researchers present results, workshops, and other activities.
- Amplified conference.
- Business conference, organized to discuss business-related matters best effected there.
- Athletic conference, a grouping of geographically-related teams.
- Conference call, in telecommunications, a "multi-party call".
- Conference hall, room where conferences are held.
- Football Conference, an English football league.
- Global meeting, example of an event or gathering about a particular subject/topic.
- News conference, an announcement to the press (print, radio, television) with the expectation of questions, about the announced matter, following.
- Settlement conference, a meeting between the plaintiff and the respondent in lawsuit, wherein they try to settle their dispute without proceeding to trial.
- Parent-teacher conference, a meeting with a child's teacher to discuss grades and school performance.
- Unconference.

ACADEMIC CONFERENCE

An academic conference is a conference for researchers (not always academics) to present and discuss their work. Together with academic or scientific journals, conferences provide an important channel for exchange of information between researchers. Generally, work is presented in the form of short, concise presentations lasting about 10 to 30 minutes, usually including discussion.

The work may be bundled in written form as academic papers and published as the conference proceedings. Often there are one or more keynote speakers (usually scholars of some standing), presenting a lecture that lasts an hour or so, and which is likely to be advertised before the conference. Panel discussions, round tables on various issues, workshops may be part of the conference, the latter ones particularly if the conference is related to the performing arts.

Prospective presenters are usually asked to submit a short abstract of their presentation, which will be reviewed before the presentation is accepted for the meeting. This peer reviewed by members of the programme committee or referees chosen by them. In some disciplines, such as English and other languages, it is common for presenters to read from a prepared script. In other disciplines such as the sciences, presenters usually base their talk around a visual presentation that displays key figures and research results. A large meeting will usually be called a conference, while a smaller is termed a workshop. They might be single track or multiple track, where the former has only one session at a time, while a multiple track meeting has several parallel sessions with speakers in separate rooms speaking at the same time.

Depending on the theme of the conference, social or entertainment activities may also be offered; if it's a large enough conference, academic publishing houses may set up displays offering books at a discount. At larger conferences, business meetings for learned societies or interest groups might also take place.

Academic conferences fall into three categories:

1. The themed conference, small conferences organized around a particular topic;
2. The general conference, a conference with a wider focus, with sessions on a wide variety of topics. These conferences are often organized by regional, national, or international learned societies, and held annually or on some other regular basis.
3. The professional conference, large conferences not limited to academics, but with academically-related issues.

Conferences are usually organized either by a scientific society or by a group of researchers with a common interest. Larger meetings may be handled on behalf of the scientific society by a Professional Conference Organiser or PCO. The meeting is announced by way of a "Call For Papers" or a Call For

Abstracts, which lists the meeting's topics and tells prospective presenters how to submit their abstracts or papers. Increasingly, submissions take place online using a managed service such as Community of Science or Oxford Abstracts.

BUSINESS CONFERENCE

Business conferences are events organized by an association, individual, publication or private company for the purpose of networking, education or to discuss a business topic with a range of speakers. They can also be organized by either a non-profit or for-profit organization. The latter is called a conference company. Business conferences are often held at convention centres and large hotels with conference facilities.

CONFERENCE CALL

A conference call is a telephone call in which the calling party wishes to have more than one called party listen in to the audio portion of the call. The conference calls may be designed to allow the called party to participate during the call, or the call may be set up so that the called party merely listens into the call and cannot speak. It is often referred to as an ATC (Audio Tele-Conference).

Conference calls can be designed so that the calling party calls the other participants and adds them to the call - however, participants are usually able to call into the conference call themselves, by dialing into a special telephone number that connects to a "conference bridge" (a specialized type of equipment that links telephone lines). Companies commonly use a specialized service provider who maintains the conference bridge, or who provides the phone numbers and PIN codes that participants dial to access the meeting or conference call. Three-way calling is available (usually at an extra charge) for many customers on their home or office phone line.

To three-way call, the first person one wishes to talk to is dialed. Then the Hook flash button (known as the recall button in the UK and elsewhere) is pressed and the other person's phone number is dialed. While it is ringing, flash/recall is pressed again to connect the three people together. This option allows callers to add a second outgoing call to an already connected call.

Usage in Business

Businesses use conference calls daily to meet with remote parties, both internally and outside of their company. Common applications are client meetings or sales presentations, project meetings and updates, regular team meetings, training classes and communication to employees who work in different locations. Conference calling is viewed as a primary means of cutting travel costs and allowing workers to be more productive by not having to go out-of-office for meetings.

Conference calls are used by nearly all United States public corporations to report their quarterly results. These calls usually allow for questions from stock analysts and are called earnings calls. A standard conference call begins with a disclaimer stating that anything said in the duration of the call may be a forward looking statement, and that results may vary significantly. The CEO, CFO, or Investor Relations officer then will read the company's quarterly report. Lastly, the call is opened for questions from analysts.

Conference calls are increasingly used in conjunction with web conferences, where presentations or documents are shared via the internet. This allows people on the call to view content such as corporate reports, sales figures and company data presented by one of the participants. The main benefit is that the presenter of the document can give clear explanations about details within the document, while others simultaneously view the presentation. Conference calls are also beginning to cross over into the world of podcasting and social networking, which in turn fosters new kinds of interaction patterns. Live streaming or broadcasting of conference calls allows a larger audience access to the call without dialing in to a bridge. In addition, organizers of conference calls can publish a dial-in number alongside the audio stream, creating potential for audience members to dial in if and when they wish to interact.

CONFERENCE HALL

A conference hall or conference room is a room provided for singular events such as business conferences. It is commonly found at large hotels and convention centres though many other establishments, including even hospitals, have one. Sometimes other rooms are modified for large conferences such as arenas or concert halls. Aircraft have been fitted out with conference rooms. Conference rooms can be windowless for security purposes. An example of one such room is in the Pentagon, known as the Tank. Typically, the facility provides furniture, overhead projectors, stage lighting, and a sound system. Smoking is normally prohibited in conference halls even when other parts of buildings permit smoking. Sometimes the term 'conference hall' is used synonymously with 'conference center' as, for example, in 'Bandaranaike Memorial International Conference Hall'.

GLOBAL MEETINGS

Global Meetings, is a list of the definitive/flagship global conferences, conventions, events, forums, meetings, summits, symposiums, trade shows, and seminars, organized by sector and topics.

In other words, these are the largest gatherings of humans around particular subjects/topics. The list is not intended to be comprehensive, but is meant to list the biggest and most popular re-occuring events. Only major topics/sectors are included and as a general rule only events with major attendance (over 100,000 per year) are included.

SETTLEMENT CONFERENCE

A settlement conference is a meeting between opposing sides of a lawsuit at which the parties attempt to reach a mutually agreeable resolution of their dispute without having to proceed to a trial.

Such a conference may be initiated through either party, usually by the conveyance of a settlement offer; or it may be ordered by the court as a precedent (preliminary step) to holding a trial. Each party, the plaintiff and the defendant, is usually represented at the settlement conference by their own Counsel or attorney.

DECISION GAMES

Decision Games are a high-impact training method to improve the decision-making and sensemaking capabilities of learners, especially in areas involving tacit knowledge that is highly subjective, ambiguous, uncertain or ill-structured. They were originally developed as "tactical decision games" by the US Marines, to accelerate the experience curve of young officers during exercises.

THEORY BEHIND DECISION GAMES

A growing body of research is revealing that the difference between experts and non-experts or novices in reacting to complex, uncertain situations lies in the way experts size-up the situation. And it is not the case of experts having access to more information than novices; on the contrary it is the case of experts sizing-up less information than novices. Malcolm Gladwell Blink, calls this capability of experts looking at less information to size-up situations as "thin-slicing" or "rapid cognition".

Klein Associates (KA), a research company based in Ohio, has done over 20 years of research on how experts size-up situations. The research has resulted in a set of knowledge elicitation techniques that can cast light on the "thin-slices" that experts pay attention to when making rapid decisions. Many studies done by KA have shown, and as one would expect, that these "thin-slices" are not usually part of the training curriculum or even in the general awareness of practitioners in an organization.

But if a knowledge-intensive organization is to survive in this hyper-competitive era, bridging this gap between what an expert sees and what a novice sees becomes extremely crucial. This is the very gap that Decision Games aim to bridge. Using KA's knowledge elicitation techniques we can get an inventory of the crucial "thin-slices" around an important decision, but how do we represent them for training? This is where narrative and screenwriting techniques come into play. Narrative techniques created by the Dave Snowden's Cynefin Centre help in using the "thin-slices" as raw material to come up with an accessible and realistic story while screenwriting techniques help in weaving real-life noise into the story.

The result is a sequence of unfolding vignettes. One Decision Game can have as many as 20-30 unfolding vignettes. The "game" part of the Decision Games is brought into play by asking participants to make a judgement on each vignette depending on the decision to be made. As this is done in a collaborative manner, participants learn from the different perspectives of their colleagues. Exposure to these different perspectives enlarges the cognitive appreciation of the learner, while the feedback they get from the game sensitises them to the significant 'thin slices' that they need to pay attention to. Decision Games are best played out in a blended mode. The value of being sensitised to different perspectives is best appreciated by learners if they play the games initially in a facilitated group setting. Once the learners are comfortable with the games and how they work, they can play another set of games in an e-learning mode. We generally recommend that a cycle of games is closed with another face to face session, so that feedback on the overall learning can be shared.

Decision Games are different from scenario-based learning in that they are:

- Based on elicited knowledge and not on readily available information.
- Based on realistic situations and they are embedded with realistic noise.
- Based on learning in a collaborative manner where the one of the primary strands of learning is by participants learning from each other's perspectives.

BASKET TRAINING

IN-BASKET TECHNIQUE

It provides trainees with a log of written text or information and requests, such as memos, messages, and reports, which would be handled by manger, engineer, reporting officer, or administrator.

PROCEDURE OF THE IN-BASKET TECHNIQUE

- In this technique, trainee is given some information about the role to be played such as, description, responsibilities, general context about the role.
- The trainee is then given the log of materials that make up the in-basket and asked to respond to materials within a particular time period.
- After all the trainees complete in-basket, a discussion with the trainer takes place.
- In this discussion the trainee describes the justification for the decisions.
- The trainer then provides feedback, reinforcing decisions made suitably or encouraging the trainee to increase alternatives for those made unsuitably.

A variation on the technique is to run multiple, simultaneous in baskets in which each trainee receives a different but organized set of information.

It is important that trainees must communicate with each other to accumulate the entire information required to make a suitable decision.

This technique focuses on:

- Building decision making skills
- Assess and develops Knowledge, Skills and Attitudes (KSAs)
- Develops of communication and interpersonal skills
- Develops procedural knowledge
- Develops strategic knowledge

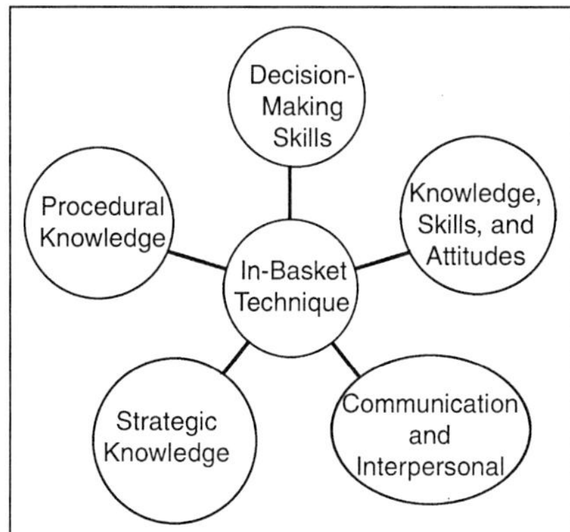

These skills are mainly cognitive to a certain extent than behavioural.

SENSITIVITY TRAINING

Sensitivity Training is a form of training that claims to make people more aware of their own prejudices, and more sensitive to others. According to its critics, it involves the use of psychological techniques with groups that its critics, *e.g.* G. Edward Griffin, claim are often identical to brainwashing tactics. Critics believe these techniques are unethical. According to his biographer, Alfred J. Marrow, Kurt Lewin laid the foundations for sensitivity training in a series of workshops he organised in 1946 to carry out a 'change' experiment, in response to a request from the Director of the Connecticut State Interracial Commission.

This led to the founding of the National Training Laboratories in Bethel, Maine in 1947. Kurt Lewin, who met Eric Trist in 1933, influenced the work of the London Tavistock Clinic, both in its work with soldiers during the second world war and in its later work with the Journal Human Relations jointly founded by a partnership of the Tavistock Institute and Lewin's group at MIT.

The nature of modern Sensitivity Training appears to be in some dispute. Its modern critics portray its origins and function in negative terms. Others view the approach as benignly beneficial in many of its historical and contemporary implementations. During World War II, Psychologists like Carl Rogers in the USA and William Sargant, John Rawlings Rees, and Eric Trist in Britain were used by the military to help soldiers deal with traumatic stress disorders (then known as Shell Shock).

This work, which required service to large numbers of patients by a small number of therapists and necessarily emphasized rapidity and effectiveness helped spur the development of group therapy as a treatment technique. Rogers and others evolved their work into new forms including encounter groups designed for persons who were not diagnosably ill but who were recognized to suffer from widespread problems associated with isolation from others common in American society. Other leaders in the development of Encounter Groups, including Will Schutz, centered their work at the Esalen Institute in Big Sur, California. Meanwhile, Training Groups or T-Groups were being developed at the National Training Labs, now part of the National Education Association.

Over time the techniques of T-Groups and Encounter Groups have merged and divided and splintered into specialized topics, seeking to promote sensitivity to others perceived as different and seemingly losing some of their original focus on self-exploration as a means to understanding and improving relations with others in a more general sense.

ORIGINS

The origins of sensitivity training can be traced as far back as 1914, when J.L. Moreno created "psychodrama," a forerunner of the group encounter (and sensitivity-training) movement. This concept was expanded on later by Kurt Lewin, a gestalt psychologist from central Europe, who is credited with organizing and leading the first T-group (training group) in 1946.

Lewin offered a summer workshop in human relations in New Britain, Connecticut. The T-group itself was formed quite by accident, when workshop participants were invited to attend a staff-planning meeting and offer feedback. The results were fruitful in helping to understand individual and group behaviour. Based on this success, Lewin and colleagues Ronald Lippitt, Leland Bradford, and Kenneth D. Benne formed the National Training Laboratories in Bethel, Maine, in 1947 and named the new process sensitivity training. Lewin's T-group was the model on which most sensitivity training at the National Training Laboratories (NTL) was based during the 1940s and early 1950s.

The focus of this first group was on the way people interact as they are becoming a group. The NTL founders' primary motivation was to help understand group processes and use the new field of group dynamics, to teach people how to function better within groups. By attending training at an offsite venue, the NTL provided a way for people to remove themselves from their everyday existence and spend two to three weeks undergoing training, thus minimizing the chances that they would immediately fall into old habits before the training truly had time to benefit its students. During this time, the NTL and other sensitivity-training programmes were new and experimental. Eventually, NTL became a nonprofit organization with headquarters in Washington, D.C. and a network of several hundred professionals across the globe, mostly based in universities. During the mid-1950s and early 1960s, sensitivity training found a place for itself, and the various methods of training were somewhat consolidated.

The T-group was firmly entrenched in the training process, variously referred to as encounter groups, human relations training, or study groups. However, the approach to sensitivity training during this time shifted from that of social psychology to clinical psychology. Training began to focus more on interpersonal interaction between individuals than on the organizational and community formation process, and with this focus took on a more therapeutic quality.

By the late 1950s, two distinct camps had been formed—those focusing on organizational skills, and those focusing on personal growth. The latter was viewed more skeptically by businesses, at least as far as profits were concerned, because it constituted a significant investment in an individual without necessarily an eye towards the good of the corporation. Thus, trainers who concentrated on vocational and organizational skills were more likely to be courted by industry for their services; sensitivity trainers more focused on personal growth were sought by individuals looking for more meaningful and enriching lives.

During the 1960s, new people and organizations joined the movement, bringing about change and expansion. The sensitivity-training movement had arrived as more than just a human relations study, but as a cultural force, in part due to the welcoming characteristics of 1960s society. This social phenomenon was able to address the unfilled needs of many members in society, and thus gained force as a social movement. The dichotomy between approaches, however, continued into the 1960s, when the organizational approach to sensitivity training continued to focus on the needs of corporate personnel.

The late 1960s and 1970s witnessed a decline in the use of sensitivity training and encounters, which had been transformed from ends in themselves into traditional therapy and training techniques, or simply phased out completely. Though no longer a movement of the scale witnessed during the 1960s, sensitivity-training programmes are still used by organizations and agencies hoping to enable members of diversified communities and workforces to better coexist and relate to each other.

GOALS OF SENSITIVITY TRAINING

According to Kurt Back, "Sensitivity training started with the discovery that intense, emotional interaction with strangers was possible. It was looked at, in its early days, as a mechanism to help reintegrate the individual man into the whole society through group development. It was caught up in the basic conflict of America at mid-century: the question of extreme freedom, release of human potential or rigid organization in the techniques developed for large combines."

The ultimate goal of the training is to have intense experiences leading to life-changing insights, at least during the training itself and briefly afterwards. Sensitivity training was initially designed as a method for teaching more effective work practices within groups and with other people, and focused on three important elements: immediate feedback, here-and-now orientation, and focus on the group process.

Personal experience within the group was also important, and sought to make people aware of themselves, how their actions affect others, and how others affect them in turn. Trainers believed it was possible to greatly decrease the number of fixed reactions that occur towards others and to achieve greater social sensitivity. Sensitivity training focuses on being sensitive to and aware of the feelings and attitudes of others.

By the late 1950s another branch of sensitivity training had been formed, placing emphasis on personal relationships and remarks. Whether a training experience will focus on group relationships or personal growth is defined by the parties involved before training begins. Most individuals who volunteer to participate and pay their own way seek more personal growth and interpersonal effectiveness. Those who represent a company, community service programme, or some other organization are more likely ready to improve their functioning within a group andor the organization sponsoring the activity. Some training programmes even customise training experiences to meet the needs of specific companies.

IN PRACTICE

An integral part of sensitivity training is the sharing, by each member of the group, of his or her own unique perceptions of everyone else present. This, in turn, reveals information about his or her own personal qualities, concerns, emotional issues, and things that he or she has in common with other members of the group.

A group's trainer refrains from acting as a group leader or lecturer, attempting instead to clarify the group processes using incidents as examples to clarify general points or provide feedback. The group action, overall, is the goal as well as the process.

Sensitivity training resembles group psychotherapy (and a technique called psychodrama) in many respects, including the exploration of emotions, personality, and relationships at an intense level. Sensitivity training, however, usually restricts its focus to issues that can be reasonably handled within the time period available. Also, sensitivity training does not include among its objectives therapy of any kind, nor does it pass off trainers/facilitators as healers of any sort. Groups usually focus on here-and-now issues; those that arise within the group setting, as opposed to issues from participants' pasts. Training does not explore the roots of behaviour or delve into deeper concepts such as subconscious motives, beliefs, etc. Sensitivity training seeks to educate its participants and lead to more constructive and beneficial behaviour. It regards insight and corrective emotional or behavioural experiences as more important goals than those of genuine therapy.

The feedback element of the training helps facilitate this because the participants in a group can identify individuals' purposes, motives, and behaviour in certain situations that arise within the group. Group members can help people to learn whether displayed behaviour is meaningful and/or effective, and the feedback loop operates continuously, extending the opportunity to learn more appropriate conduct. Another primary principle of sensitivity training is that of feedback; the breakdown of inhibitions against socially repressed assertion such as frankness and self-expression are expected in place of diplomacy.

Encounters that take place during sensitivity training serve to help people practice interpersonal relations to which they are likely not accustomed. The purpose is to help people develop a genuine closeness to each other in a relatively short period of time. Training encounters are not expected to take place without difficulty. Many trainers view the encounter as a confrontation, in which two people meet to see things through each other's eyes and to relate to each other through mutual understanding.

There is a difference between the scientific study of group dynamics (a branch of social psychology) and the human relations/group workshop aspect. The popularity of sensitivity training during the 1960s was due in large part to the emotional, experiential aspect. Yet many pragmatic advocates of sensitivity training felt it was necessary to avoid working with the most emotional converts, and conducted experiments in a laboratory in as realistic a situation as could be approxi-mated, seeking a scientific approach more characteristic of psychological studies.

Other programmes, not so concerned with the scientific validity of their studies or with freedom from distraction, offer full-time training programmes during the day. Participants can choose on their own whether or not to maintain contact with the office for the duration of training. Others offer part-time sessions for several hours a day, and the participants' daily routine is otherwise uninterrupted. Sensitivity-training programmes generally last a few days, but some last as many as several weeks.

T-GROUPS

Within most training groups (T-groups), eight to ten people meet with no formal leader, agenda, or books-only a somewhat passive trainer. Trainers do not necessarily direct progress, just help participants to understand what is happening within the group. In defining a T-group, Robert T. Golembiewski explains the major distinguishing features as follows: "it is a learning laboratory; it focuses on learning how to learn; and it distinctively does so via a 'here-and-now' emphasis on immediate ideas, feelings, and reactions." The learning takes place within a group's struggle to create something meaningful for itself in an essentially unstructured setting.

Issues that traditionally arise in such a setting include developing group norms and cohesion, reasons for scape-goating, selective communication channels, struggles for leadership, and collective decision-making patterns. Power struggles and decision-making conflict are the most prevalent problems as groups work towards establishing an identity and meet individual member needs. More specifically, group members can help each other identify when they are: attempting to control others or, conversely, when they are seeking support; punishing themselves or other group members; withdrawing from the group; trying to change people rather than accepting them; reacting emotionally to a given situation; and ignoring, rather than scrutinizing, behaviour between group members.

Ultimately, T-groups were not a tremendously successful part of the sensitivity-training movement. This was in part because T-group trainers do not actually teach, but help people learn by assuming a more passive role. This sometimes confuses and upsets those who expect and desire more guidance. Another reason is that despite the intensity of the learning experience, most participants have difficulty quantifying exactly what they have learned and why it matters.

IN ORGANIZATIONAL

Organizational goals appear to be the antithesis of those of sensitivity training. Sensitivity training is fueled on emotional outbursts in group settings, possibly leading to a change in attitude towards another individual. Desired results include more openness, spontaneity, and sensitivity to others.

And while organizations are made up of people who interact and could benefit from such training, the goals of an organization are often more related to increased production or higher profit margins than modifying means of interpersonal communication. To make sensitivity training work in organizational settings, the training must be adapted to the goals of the particular organization.

In its orientation as a study of group dynamics, sensitivity training is similar to the general concept of organizational development, a process by which organizations educate themselves in order to achieve better problem-solving capabilities. However, most sensitivity programmes do focus on individual behaviour within groups, while organizational development focuses on the group and how it works as a whole. Also, sensitivity-training groups are often composed entirely of people who are strangers to each other, while organizational-development programmes seek to educate groups of people with shared working histories and experiences. Finally, the end goals of these training programmes differ significantly. Sensitivity training, if successful, leads to self-awareness and insight that will help its participants in all aspects of life (including the work-place).

Organizational development places more of its focus on becoming aware of one's role within work-place dynamics, leading to more effective group functioning (one of sensitivity training's goals, but with a more defined group in which to function).

POLITICAL CORRECTNESS AND THE RESPONSE TO SENSITIVITY TRAINING

The development of sensitivity training has led many critics to claim that such training is not really designed to help people be more sensitive to other people's ideas and feelings, but it is really crafted to change one's attitudes, standards and beliefs. These critics argue that sensitivity training merely wears people down until they conform to the mentality of the group, and agree that views of the group are acceptable, regardless of the value of the group idea or belief.

These critics further assert that sensitivity training is often misused to force people into complying with community directives to conform to standards of political correctness. Political correctness has been defined as "avoidance of expressions or actions that can be perceived to exclude or marginalize or insult people who are socially disadvantaged or discriminated against" or the "alteration of language to redress real or alleged injustices and discrimination or to avoid offense."

For example, the politically correct (PC) word for someone who is crippled would be disabled, and the PC word for someone who is blind would be visually impaired. While political correctness seems like a good thing, opponents of the political correctness movement argue that it represents a totalitarian movement towards an ideological state in which citizens will be terrorized into conforming with the PC movement or risk punishment by the State. This friction between advocates for sensitivity training and opponents of the PC movement has resulted in an emotional reaction to sensitivity training the workplace. In spring, 2000, the Environmental Protection Agency announced to its Washington-area employees that it was planning a series of sensitivity training seminars to "create understanding, sensitivity and awareness of diversity issues and provide a forum for exchanging information and ideas." The course failed miserably. The EPA employees complained the course literature was condescending and one-sided. Many employees seemingly felt that only certain ones of them were being asked to be sensitive to the others. Proponents of the PC movement assert that it merely makes each of us a bit more sensitive to the challenges that our fellow citizens may face on a day-by-day basis. Clearly, the debate will continue. Sensitivity training will continue, and employers and other organizations will continue to assess whether its effectiveness warrants the costs.

HUMAN RELATIONS TRAINING

The human relations training programme was created to assess the effects of human relations training on managerial effectiveness. The training consisted of 28-weekly, 90-minute sessions. Divided into phases, phase I of the training focused on discussions of leaders, leadership, followership, and leadership styles.

Phase II, the largest component of the training, was devoted to experiential learning exercises such as self-ratings on the managerial grid, partition exercises, judgement, in-basket, listening and interview exercises. Phase III of the training focused on motivation theories. Pre and post measures of self-awareness, sensitivity to the needs of others, and leadership styles were completed. Behaviour ratings by supervisors and subordinates were also collected. Evaluation was done 90 days and 18 months after completion of the programme.

The training was found to be effective in changing attitudes and behaviours and these changes were related to increased managerial effectiveness. Managers who completed the training increased their self-awareness, were more sensitive to the needs of others, and focused on developing mutual trust with their employees. Subordinates reported that they had better rapport and communication with managers who had completed the training. The first training in emotional and social competence in organizations was called "human relations training".

Between 1950 and 1975 there were hundreds of human relations training programmes offered to thousands of managers in American organizations. Most of these efforts were not evaluated, and many were disappointing in their lack of lasting impact. However, one such programme stands out as an exception. The managerial human relations training programme described by Hand, Richards, and Slocum (1973) was developed at the Pennsylvania State University. It was an "off the shelf" programme conducted by the Continuing Education Division.

Through that division, it was delivered numerous times in firms throughout a several state area. The objective of the programme was to encourage participants to utilize human relations principles in their dealings with employees. More specifically, the programme designers sought to increase participants' use of consideration and initiating structure. Thus, the programme targeted the emotional competencies of self-awareness, empathy and leadership. The training consisted of 90-minute sessions given once a week for 28 weeks.

The first phase of the training was devoted to a discussion of managerial styles. Topics included leaders, leadership, and followership; initiating structure and consideration; and autocratic, democratic, and laissez-faire styles of management. This first phase, which involved primarily cognitive learning, lasted approximately 9 hours. The second phase of the training was primarily experiential in nature. There were numerous individual and group exercises including self-ratings on the managerial grid, an in-basket exercise, a listening exercise, and a corrective interview role play. In each exercise, an individual or group would perform a task while the rest of the participants observed them. Following completion of the task, the observers would give critical feedback on what they observed.

All the participants then would discuss the feedback. Thirty hours were devoted to this experiential learning. The final phase of the programme was devoted to discussion of the motivational theories of Porter, McGregor, Herzberg, and Maslow. It lasted about three hours. The total programme involved 42 hours of training. A rigorous evaluation study of the programme occurred when it was implemented in a specialty steel plant located in central Pennsylvania. The design involved both a trained group of managers and a control group that did not receive training.

It also involved pre-training, post-training, and long-term follow-up measures of managerial attitudes, leadership behaviour as perceived by subordinates, and performance as rated by superiors. The post-training measures were completed 90, days following training, and the long-term follow-up assessment occurred 18 months after the completion of training.

The results indicated no differences between the two groups at the 90-day post-training assessment, but there were several significant differences at the 18-month follow-up. By that time, the trained managers had become significantly more self-aware and more sensitive to the needs of others in their attitudes.

Their subordinates also perceived them as having improved in rapport and two-way communication. The controls, on the other hand, did not change in their attitudes, and their subordinates perceived them as significantly less considerate than they had been at the time of the pre-training assessment. Performance ratings also improved for trained managers working in consultative climates, but not for those working in autocratic climates. In contrast, performance ratings for untrained controls declined over time.

TRANSACTIONAL ANALYSIS

Transactional analysis is a social psychology developed by Eric Berne, MD. Over the past four decades Eric Berne's theory has evolved to include applications to psychotherapy, counselling, education, and organizational development.

PSYCHOTHERAPY

Transactional analysis is a powerful tool to bring about human well-being. In psychotherapy, transactional analysis utilizes a contract for specific changes desired by the client and involves the "Adult" in both the client and the clinician to sort out behaviours, emotions and thoughts that prevent the development of full human potential. Transactional analysts intervene as they work with clients in a safe, protective, mutually respectful-OK/OK— environment to eliminate dysfunctional behaviours and establish and reinforce positive relationship styles and healthy functioning. Transactional analysts are able to use the many tools of psychotherapy, ranging from psychodynamic to cognitive behavioural methods in effective and potent ways. Examples of transactional analysis psychotherapy can bee seen in our Master Therapists series, the Ellyn Bader and Peter Pearson Couples Therapy Videotapes and the Carlo Moiso-Isabelle Crespelle DVD.

COUNSELLING

Counselors who utilize transactional analysis work contractually on solving "here and now" problems. Counselling work focuses on creating productive problem solving behaviours.

Using transactional analysis, counselor's establish an egalitarian, safe and mutually respectful working relationship with their clients. This working relationship provides tools clients can utilize in their day-to-day functions to improve the quality of their lives.

EDUCATIONAL

Transactional Analysis is a practical educational psychology that offers a way of transforming educational philosophy and principles into everyday practice. TA concepts provide a flexible and creative approach to understanding how people function and to the connections between human behaviour, learning and education.

Teaching them to both teachers and students is a process of empowerment, enhancing effective methods of interaction and mutual recognition. Educational TA is both preventive and restorative. TA concepts are developed and used with people of all ages and stages of development in their various social settings.

The aim is to increase personal autonomy, to support people in developing their own personal and professional philosophies and to enable optimum psychological health and growth.

The key philosophical concepts that underpin Educational TA are:

- Effective educators offer empathic acceptance of all human beings as people together with respect for their dignity. These qualities are at the heart of successful learning relationships.
- People at any age and stage can learn to take responsibility for their own decisions and actions.
- Educational difficulties can be addressed effectively with co-operative goodwill and a coherent theoretical framework that makes sense of the human dynamics involved.

The process of educational TA is contractual, so that all parties know where they stand, and what agreements have been made for what purposes. Throughout the process the ideas and methods of TA are used openly to promote informed co-operation and the sharing of power between all parties.

TA can be used to address important issues in:

- initial and continuing teacher education
- institutional climate and culture
- developmental and educational needs
- self esteem building
- parent education
- student motivation
- staff morale and teacher well-being
- blocks to learning and teaching
- behaviour management

All educational TA is invaluable in helping people to thrive and in promoting healthy and effective learning in a wide variety of contexts.

ORGANIZATIONAL

Transactional Analysis is a powerful tool in the hands of organizational development specialists. Through presenting the basic concepts of transactional analysis and using it as the basic theory to undergird the objectives of their clients, organizational development specialists build a common strategy with which to address the particular needs of organizations and to build a functional relationship, as well as eliminate dysfunctional organizational behaviours.

3

Leadership

DEFINITION OF LEADERSHIP

Leadership is stated as the "process of social influence in which one person can enlist the aid and support of others in the accomplishment of a common task." Definitions more inclusive of followers have also emerged. Alan Keith stated that, "Leadership is ultimately about creating a way for people to contribute to making something extraordinary happen." Tom DeMarco says that leadership needs to be distinguished from posturing.

The following parts discuss several important aspects of leadership including a description of what leadership is and a description of several popular theories and styles of leadership. This substance also discusses topics such as the role of emotions and vision, as well as leadership effectiveness and performance, leadership in different contexts, how it may differ from related concepts (*i.e.*, management), and some critiques of leadership as generally conceived.

LEADERSHIP STYLES

AUTOCRATIC LEADERSHIP

Autocratic leadership is a classical leadership style with the following characteristics:

- Manager seeks to make as many decisions as possible.
- Manager seeks to have the most authority and control in decision making.
- Manager seeks to retain responsibility rather than utilise complete delegation.
- Consultation with other colleagues in minimal and decision-making becomes a solitary process.
- Managers are less concerned with investing their own leadership development, and prefer to simply work on the task at hand.

The autocratic leadership style is seen as an old fashioned technique. It has existed as long as managers have commanded subordinates, and is still employed by many leaders across the globe. The reason autocratic leadership survives, even if it is outdated, is because it is intuitive, carries instant benefits, and comes natural to many leaders. Many leaders who start pursuing leadership development are often trying to improve upon their organisations autocratic leadership style.

What are the Benefits of the Autocratic Leadership Style?

Despite having many critics, the autocratic leadership styles offer many advantages to managers who use them.

These include:

- Reduced stress due to increased control. Where the manager ultimately has significant legal and personal responsibility for a project, it will comfort them and reduce their stress levels to know that they have control over their fate.
- A more productive group 'while the leader is watching'. The oversight that an autocratic manager exerts over a team improves their working speed and makes them less likely to slack. This is ideal for poorly motivated employees who have little concern or interest in the quality or speed of work performed.
- Improved logistics of operations. Having one leader with heavy involvement in many areas makes it more likely that problems are spotted in advance and deadlines met. This makes autocratic leadership ideal for one-off projects with tight deadlines, or complicated work environments where efficient cooperation is key to success.
- Faster decision making. When only one person makes decisions with minimal consultation, decisions are made quicker, which will allow the management team to respond to changes in the business environment more quickly.

What are the Disadvantages of the Autocratic Leadership Style?

Short-termistic Approach to Management

While leading autocratically will enable faster decisions to be made in the short term, by robbing subordinates of the opportunity to gain experience and start on their own leadership development, and learn from their mistakes, the manager is actually de-skilling their workforce which will lead to poorer decisions and productivity in the long run.

Manager Perceived as Having Poor Leadership Skills

While the autocratic style has merits when used in certain environments, autocratic leadership style is easy yet unpopular. Managers with poor leadership skills with often revert to this style by default.

Increased Workload for the Manager

By taking on as much responsibility and involvement as possible, an autocratic leader naturally works at their full capacity, which can lead to long term stress and health problems and could damage working relationships with colleagues. This hyper-focus on work comes at the expense of good leadership development.

People Dislike being Ordered Around

They also dislike being shown very little trust and faith. As a result, the autocratic leadership style can result in a demotivated workforce. This results in the paradox that autocratic leadership styles are a good solution for demotivated workers, but in many cases, it is the leadership style alone that demotivates them in the first place. Generation/employees particularly dislike this style.

Teams become Dependent upon their Leader

After becoming conditioned to receive orders and act upon them perfectly, workers lose initiative and the confidence to make decisions on their own. This results in teams of workers who become useless at running operations if they loose contact with their leader. This is the result of a lack of time dedicated to leadership development on the employees part.

When is the Autocratic Leadership Style Effective?

Following on from the merits and drawbacks, the autocratic leadership style is useful in the following work situations:

- Short term projects with a highly technical, complex or risky element.
- Work environments where spans of control are wide and hence the manager has little time to devote to each employee.
- Industries where employees need to perform low-skilled, monotonous and repetitive tasks and generally have low levels of motivation.
- Projects where the work performed needs to be completed to exact specifications and/or with a tight deadline.
- Companies that suffer from a high employee turnover, *i.e.* where time and resources devoted to leadership development would be largely wasted. Although one could argue that a lack of leadership development in the first place caused the high turnover.

DEMOCRATIC LEADERSHIP

Democratic Leadership is the leadership style that promotes the sharing of responsibility, the exercise of delegation and continual consultation.

The style has the following characteristics:

- Manager seeks consultation on all major issues and decisions.
- Manager effectively delegate tasks to subordinates and give them full control and responsibility for those tasks.

- Manager welcomes feedback on the results of intiatives and the work environment.
- Manager encourages others to become leaders and be involved in leadership development.

What are the Benefits of the Democratic Leadership Style?

Positive Work Environment

A culture where junior employees are given fair amount of responsibility and are allowed to challenge themselves is one where employees are more enthused to work and enjoy what they do. Successful initiatives. The process of consultation and feedback naturally results in better decision-making and more effective operations.

Companies run under democratic leadership tend to run into fewer grave mistake and catastrophes. To put it simply —people tell a democratic leader when something is going badly wrong, while employees are encouraged to simply hide it from an autocrat.

Creative Thinking

The free flow of ideas and positive work environment is the perfect catalyst for creative thinking.

The benefits of this aren't just relevant for creative industries, because creative thinking is required to solve problems in every single organisation, whatever it's nature. Reduction of friction and office politics. By allowing subordinates to use their ideas and even more importantly—gain credit for them, you are neatly reducing the amount of tension employees generate with their manager.

When autocratic leaders refuse to listen to their workers, or blatantly ignore their ideas, they are effectively asking for people to talk behind their back and attempt to undermine or supercede them.

Reduced Employee Turnover

When employees feel empowered through leadership development, a company will experience lower rates of employee turnover which has numerous benefits.

A company that invests in leadership development for its employees, is investing in their future, and this is appreciated by a large majority of the workforce.

What are the Disadvantages of Democratic Leadership Style? Lengthy and 'Boring' decision-making.

Seeking consultation over every decision can lead to a process so slow that it can cause opportunities to be missed, or hazards avoided too late.

Danger of Pseudo Participation

Many managers simply pretend to follow a democratic leadership style simply to score a point in the eyes of their subordinates. Employees are quick to realise when their ideas aren't actually valued, and that the manager is merely following procedure in asking for suggestions, but never actually implementing them. In other words, they're simply exerting autocratic leadership in disguise.

When is the Democratic Leadership Style Effective?

Now you've heard about the benefits and drawbacks of this leadership style, let's look at where its actually implemented in the business world. (1) Democratic leadership is applied to an extent in the manufacturing industry, to allow employees to give their ideas on how processes can become leaner and more efficient.

While 'Fordism' is still applied in some factories across the country, truth is that production managers are now really starting to harness the motivational bonuses associated with not treating employees like robots anymore. (2) Democratic leadershp is effective in proffessional organisations where the emphasis is clearly on training, professional and leadership development and quality of work performed.

Democratic procedures are simply just one cog in the effective leadership mechanisms firms like The Big Four have created over the years. (3) Non profit organisations also tremendously benefit from drawing upon the creative energies of all their staff to bring about cost cutting techniques or fund raising ideas.

Ceative industries such as advertising and television enjoy alot of benefits from the free flow of ideas that democratic leadership brings.

BUREAUCRATIC LEADERSHIP

The bureaucratic leadership style is concerned with ensuring workers follow rules and procedures accurately and consistently.

Bureaucratic leadership normally has the following characteristics:

- Leaders expect a employees to display a formal, business-like attitude in the workplace and between each other.
- Managers gain instant authority with their position, because rules demand that employees pay them certain priveledges, such as being able to sign off on all major decisions. As a result, leaders suffer from 'position power'. Leadership development becomes pointless, because only titles and roles provide any real control or power.
- Employees are rewarded for their ability to adhere to the rules and follow procedure perfectly.
- Bureaucratic systems usually gradually develop over a long period of time, and hence are more commonly found in large and old businesses.

What are the Benefits of the Bureaucratic Leadership Style?

Increased safety

In dangerous workplaces where procedures save lives, a bureaucratic management style can help enforce health and safety rules.

Quality work

Some tasks, such as completing proffessional work or medical examinations, need to be done in a meticulous fashion to be done correctly. Laziness can result in poor work, and hence one solution is to enforce the rules via the bureacratic leadership style.

Ultimate control

An environment whereby employees are intrinsically motivated to follow rules in order to be promoted and succeed results in the tightest control management can ever assume over a company. This control can be used to cut costs or improve productivity.

What are the Disadvantages of the Bureaucratic Leadership Style?

Dehumanises the Business

Bureaucratic companies tend to remove as much potential for 'human error' out of the picture as possible. Unfortunately this also has the effect of removing all the enjoyment and reward that comes from deciding how to do a task and accomplishing it.

Lack of Self-fulfillment

The bureaucratic way of working hampers employees efforts to become successful and independent, because the system becomes too contraining.

Parkinson's Law

Cyril Northcote Parkinson made the scientific observation that the number of staff in bureaucracies increased by an average of 5 per cent-7 per cent per year "irrespective of any variation in the amount of work (if any) to be done."".

He explains this growth by two forces:

1. "An official wants to multiply subordinates, not rivals
2. "Officials make work for each other."

Parkinson's findings suggest that bureaucratic leadership encourages inefficiency and waste of internal resources in the long run.

'Position Power' Obessession

After working in an environment that reinforces the idea that authority is created by rules which in turn support senior positions. Employees become attached to the idea that simply being in a job position creates authority. This can lead to intense office politics, arrogant leaders and little incentive to perform well once an employee has landed a top job.

Lack of Creativity

It goes without saying that a rule-based culture hinders creativity and encourages workers to simply perform puppet-like work rather than think independently. This may result in a lack of growth in the business due to employees simply not thinking out of the box or looking for new areas to develop.

Poor Communication

A common feature of a bureaucratic system is a complicated network of communication lines. Managers who don't want to be 'bothered' by junior staff simply create procedures that allow them to avoid communicating with those below them.

'Go through the formal process', 'Talk to my secretary' and 'My schedule is full' are common rule-based excuses for blocked contact. Barriers to communication can hinder the success of any company. For example, the board may be charging ahead with a doomed product simply because their shop floor workers cannot pass on the message that customers are giving very negative feedback.

When is the Bureaucratic Leadership Style Effective?

Bureaucratic leadership is found in extremely large corporations such as General Electric, Daimler and General Motors. However these cultures have evolved due to the age and size of these companies, and are generally blamed for the slow growth and recent failures at these companies. (1) Governmental bodies often have bureaucratic systems, and while these are often despised by the public, they ensure accountability to the tax payer and fair treatment for all. Excessive form-filling also serves the purpose of passing effort from the government authority (with a tight budget) onto the individual, helping to save costs. (2) Dangerous workplaces such as mines, oil rigs, construction sites and film sets all benefit from the tight control over health and safety that rules offer.

OTHER LEADERSHIP STYLES

What Different Leadership Types are there?

These 3 key management leadership styles are by no means a comprehensive list. Different leadership styles include *laissez-faire* leadership, where the leader sets tasks and leaves workers up to their own devices to complete it. To help you discover which leadership style you possess, try our new Leadership Style Questionnaire.

Leadership Development

Leadership development is a complicated area, and thus countless styles have been theorised and researched. Good leadership development often involves using resources such as Leadership Expert to be 'sift' through these different leadership development tips and ideas. Once you've been able to pull together a solid leadership development plan for yourself, you can start to really engage your employees—and maybe even set them off on their own leadership development quest! To further their leadership education, people often check the online MBA rankings to see whether MBA online is a match for them. Leadership Expert also has substances on another leadership style: charismatic leadership. You may also want to read our substances on leadership theories and common leadership traits.

POSITIVE LEADER

At Positive Leadership Solutions, positive psychological principles are used to help leaders, teams and organizations successfully meet goals, increase profitability, and enhance effectiveness. By focusing on strengths, people take responsibility for creating exceptional teams and organizations. Central to understanding positive leadership is the positive psychology movement.

Positive Psychology is a branch of psychology that uses scientific principles to understand what makes people healthy and helps them thrive. Positive Psychologists research positive emotions such as happiness or hope and positive individual traits such as wisdom or social intelligence. They might also focus on how groups or organizations build better communities and healthy places to work; often referred to as Positive Organizational Scholarship or Positive Organizational Behaviour. A central element of Positive Psychology is that it is research-based, not simply a set of "feel good" ideas.

POSITIVE LEADERSHIP FOCUSES ON RELATIONSHIPS, NOT ROLES

Every person has the potential to be a positive force in the workplace or community. Positive Leaders consistently match their values and beliefs to building something that improves their world. A high tide raises all boats.

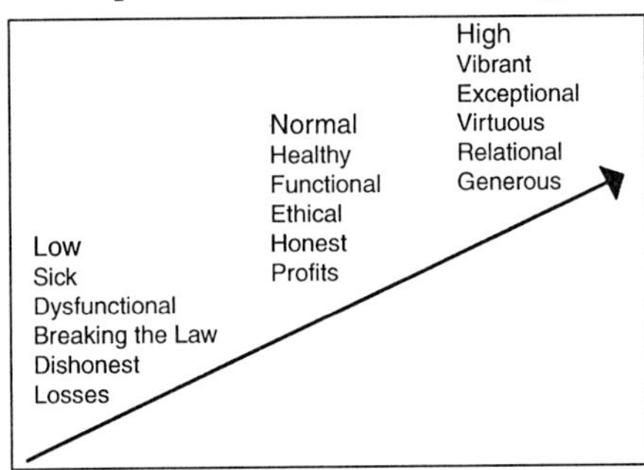

People in organizations can work towards building a truly exceptional workplace, but often miss the opportunity. This oversight costs the organization, the employees and the community. Through assessment, self-reflection and executive coaching, you can learn to improve and create a workplace that attracts great employees, managers and clients.

CHARACTERISTICS OF A GREAT LEADER

There is a saying that great leaders are made not born. This should give you hope that once you understand the characteristics of a great leader you can become one yourself. Here are a few things that every great leader does or possesses within themselves to inspire people around them.

- Leaders are always improving. They understand that things are changing around them and for them to be great leader they must be changing too!
- Great leaders inspire people around them to become better. People want to do their best because of their leadership.
- Leaders know how to concentrate on people's strengths and not their weaknesses. Everyone has things that they are good at and leaders know how to bring this out in a person.
- Leaders are pro active and not reactive. They understand the importance of leading their people and not waiting for somebody else to get started.
- Leaders treat people with respect and importance. They know that to get a person to do something they need to want to do it. Treating people with respect and importance is a technique that works for this.
- Great leaders are self-motivated. They understand there will be ups and downs in their life and in their business, but they stay positive and do not let outside influences affect their attitudes.
- Leaders are well spoken. They know how to say the right thing at the right time.
- Leaders are always prepared. They don't leave things to chance but rather they control situations through preparation.
- Great leaders do not have big egos. They care about others as opposed to being a self centered individual.
- Great leaders are great mentors. They know how to pass on the knowledge that they themselves have personally already attained.
- Leaders are people who write down goals and strive to achieve them. They understand the importance of goal setting and the example that they are teaching by doing this.
- Great leaders are ambitious hard workers. They never expect more out of the people around them then they are willing to give themselves. This type of attitude is contagious and leads to everybody working harder.

NEGATIVE LEADER

Negative leaders act domineering and superior with people. They believe the only way to get things done is through penalties, such as loss of job, days off without pay, reprimanding employees in front of others, etc. They believe their authority is increased by frightening everyone into higher levels of productivity. Yet what always happens when this approach is used wrongly is that morale falls; which of course leads to lower productivity.

Also note that most leaders do not strictly use one or another, but are somewhere on a continuum ranging from extremely positive to extremely negative. People who continuously work out of the negative are bosses while those who primarily work out of the positive are considered real leaders.

4

Motivation

Motivation is the activation or energization of goal-orientated behaviour. Motivation is said to be intrinsic or extrinsic. The term is generally used for humans but, theoretically, it can also be used to describe the causes for animal behaviour as well. This substance refers to human motivation. According to various theories, motivation may be rooted in the basic need to minimize physical pain and maximize pleasure, or it may include specific needs such as eating and resting, or a desired object, hobby, goal, state of being, ideal, or it may be attributed to less-apparent reasons such as altruism, selfishness, morality, or avoiding mortality. Conceptually, motivation should not be confused with either volition or optimism. Motivation is related to, but distinct from, emotion.

Intrinsic motivation refers to motivation that is driven by an interest or enjoyment in the task itself, and exists within the individual rather than relying on any external pressure. Intrinsic motivation has been studied by social and educational psychologists since the early 1970s. Research has found that it is usually associated with high educational achievement and enjoyment by students. Intrinsic motivation has been explained by Fritz Heider's attribution theory, Bandura's work on self-efficacy, and Deci and Ryan's cognitive evaluation theory.

Students are likely to be intrinsically motivated if they:

- Attribute their educational results to internal factors that they can control (*e.g.* the amount of effort they put in),
- Believe they can be effective agents in reaching desired goals (*i.e.* the results are not determined by luck), are interested in mastering a topic, rather than just rote-learning to achieve good grades.

Extrinsic motivation comes from outside of the individual. Common extrinsic motivations are rewards like money and grades, coercion and threat of punishment. Competition is in general extrinsic because it encourages the performer to win and beat others, not to enjoy the intrinsic rewards of the activity.

A crowd cheering on the individual and trophies are also extrinsic incentives. Social psychological research has indicated that extrinsic rewards can lead to overjustification and a subsequent reduction in intrinsic motivation. In one study demonstrating this effect, children who expected to be (and were) rewarded with a ribbon and a gold star for drawing pictures spent less time playing with the drawing materials in subsequent observations than children who were assigned to an unexpected reward condition and to children who received no extrinsic reward.

Self-determination theory proposes that extrinsic motivation can be internalised by the individual if the task fits with their values and beliefs and therefore helps to fulfill their basic psychological needs. Internalised extrinsic motivation has been shown to lead to more positive outcomes, such as wellbeing, increased productivity and task satisfaction. The self-control of motivation is increasingly understood as a subset of emotional intelligence; a person may be highly intelligent according to a more conservative definition (as measured by many intelligence tests), yet unmotivated to dedicate this intelligence to certain tasks.

Yale School of Management professor Victor Vroom's "expectancy theory" provides an account of when people will decide whether to exert self control to pursue a particular goal. Drives and desires can be described as a deficiency or need that activates behaviour that is aimed at a goal or an incentive. These are thought to originate within the individual and may not require external stimuli to encourage the behaviour.

Basic drives could be sparked by deficiencies such as hunger, which motivates a person to seek food; whereas more subtle drives might be the desire for praise and approval, which motivates a person to behave in a manner pleasing to others.

By contrast, the role of extrinsic rewards and stimuli can be seen in the example of training animals by giving them treats when they perform a trick correctly. The treat motivates the animals to perform the trick consistently, even later when the treat is removed from the process.

MODEL OF MOTIVATION

Over the years many psychologists have attempted to define and categorise what motivates people. This became particularly important after the Second World War as the Western nations attempted to rebuild their drained industrial economies, and during the 50s and 60s much was researched and written about Human Relations.

It was recognised that people who worked in organisations were more than just numbers and, if properly managed, could not only produce more, but also contribute more. This is not the place to cover the work of every motivational theorist: we've simply chosen a couple that have entered the mainstream management vocabulary:

- Theory X/Theory Y
- Herzberg's Motivation - Hygiene Theory

THEORY X/THEORY Y

Douglas McGregor published "The Human Side of Enterprise" in 1960, in which he suggested that traditional management methods (which he called Theory X) might not be the only way to get people motivated. Instead, you could take a different approach (based on Theory Y) and achieve the same if not more.

Theory X is the traditional view of direction and control, based on these assumptions:

- The average person inherently dislikes work and will avoid it if at all possible.
- As a result, most people have to be coerced, controlled and threatened if they are to put in enough effort to achieve the organisation's goals.
- In fact the average person prefers to be directed, avoids responsibility, isn't ambitious and simply seeks security.

Theory Y, based on the integration of individual and organisational goals, assumes:

- The physical and mental effort of work is as natural as play or rest, so the average person doesn't inherently dislike work.
- We are capable of self-direction and self-control, so those factors don't necessarily have to come from elsewhere.
- Our commitment to an objective is a function of the rewards for its achievement.
- The average person learns not only to accept but to seek responsibility.
- Most people have a capacity for imagination, ingenuity and creativity.
- The intellectual potential of most people is under-used in modern industrial life.

Theory Y is not a soft option. In fact it can take as much management effort as Theory X, but the effects of a Theory Y approach will last longer. The Theory X manager is a dying breed, and Theory Y lies behind most modern approaches to motivation. Nowadays the terminology is used as a polite way of referring to the old command-and-control approach to management: the trouble is the diehard Theory X manager won't pick up the subtle criticism!

HERZBERG'S MOTIVATION - HYGIENE THEORY

Frederick Herzberg studied and practised clinical psychology in Pittsburgh, where he researched the work-related motivations of thousands of employees. His findings were published in "The Motivation to Work" in 1959. He concluded that there were two types of motivation: Hygiene Factors that can demotivate if they are not present - such as supervision, interpersonal relations, physical working conditions, and salary. Hygiene Factors affect the level of dissatisfaction, but are rarely quoted as creators of job satisfaction. Motivation

Factors that will motivate if they are present-such as achievement, advancement, recognition and responsibility. Dissatisfaction isn't normally blamed on Motivation Factors, but they are cited as the cause of job satisfaction.

So, once you've satisfied the Hygiene factors, providing more of them won't generate much more motivation, but lack of the Motivation Factors won't of themselves demotivate. There are clear relationships to Maslow here, but Herzberg's ideas really shaped modern thinking about reward and recognition in major companies.

ACHIEVEMENT MOTIVATION

Over the years, behavioural scientists have noticed that some people have an intense desire to achieve something, while others may not seem that concerned about their achievements. This phenomenon has attracted a lot of discussions and debates. Scientists have observed that people with a high level of achievement motivation exhibit certain characteristics. Achievement motivation is the tendency to endeavor for success and to choose goal oriented success or failure activities. Achievement motivation forms to be the basic for a good life. People who are oriented towards achievement, in general, enjoy life and feel in control. Being motivated keeps people dynamic and gives them self-respect.

They set moderately difficult but easily achievable targets, which help them, achieve their objectives. They do not set up extremely difficult or extremely easy targets. By doing this they ensure that they only undertake tasks that can be achieved by them. Achievement motivated people prefer to work on a problem rather than leaving the outcome to chance. It is also seen that achievement motivated people seem to be more concerned with their personal achievement rather than the rewards of success. It is generally seen that achievement motivated people evidenced a significantly higher rate of advancement in their company compared to others.

Programmes and courses designed, involves seven "training inputs." The *first* step refers to the process through which achievement motivation thinking is taught to the person. The *second* step helps participants understand their own individuality and goals.

The *third* assist participants in practicing achievement-related actions in cases, role-plays, and real life. A *fourth* refers to practicing of achievement-related actions in business and other games. A *fifth* input encourages participants to relate the achievement behaviour model to their own behaviour, self-image, and goals.

The *sixth* programme facilitates participants to develop a personal plan of action. *Finally*, the course provides participants with feedback on their progress towards achieving objectives and targets. Achievement motivation as a branch of study has greatly established its prominence. A number of companies are now training their employees in the same.

EFFECTS OF ACHIEVEMENT MOTIVATION ON BEHAVIOUR

Motivation is the basic drive for all of our actions. Motivation refers to the dynamics of our behaviour, which involves our needs, desires, and ambitions in life. Achievement motivation is based on reaching success and achieving all of our aspirations in life.

Achievement goals can affect the way a person performs a task and represent a desire to show competence. These basic physiological motivational drives affect our natural behaviour in different environments. Most of our goals are incentive-based and can vary from basic hunger to the need for love and the establishment of mature sexual relationships. Our motives for achievement can range from biological needs to satisfying creative desires or realizing success in competitive ventures. Motivation is important because it affects our lives everyday. All of our behaviours, actions, thoughts, and beliefs are influenced by our inner drive to succeed.

Implicit and Self-Attributed Motives

Motivational researchers share the view that achievement behaviour is an interaction between situational variables and the individual subject's motivation to achieve. Two motives are directly involved in the prediction of behaviour, implicit and explicit. Implicit motives are spontaneous impulses to act, also known as task performances, and are aroused through incentives inherent to the task.

Explicit motives are expressed through deliberate choices and more often stimulated for extrinsic reasons. Also, individuals with strong implicit needs to achieve goals set higher internal standards, whereas others tend to adhere to the societal norms. These two motives often work together to determine the behaviour of the individual in direction and passion. Explicit and implicit motivations have a compelling impact on behaviour. Task behaviours are accelerated in the face of a challenge through implicit motivation, making performing a task in the most effective manner the primary goal. A person with a strong implicit drive will feel pleasure from achieving a goal in the most efficient way.

The increase in effort and overcoming the challenge by mastering the task satisfies the individual. However, the explicit motives are built around a person's self-image.

This type of motivation shapes a person's behaviour based on their own self-view and can influence their choices and responses from outside cues. The primary agent for this type of motivation is perception or perceived ability. Many theorists still cannot agree whether achievement is based on mastering one's skills or striving to promote a better self-image. Most research is still unable to determine whether these different types of motivation would result in different behaviours in the same environment.

The Hierarchal Model of Achievement Motivation

Achievement motivation has been conceptualized in many different ways. Our understanding of achievement-relevant effects, cognition, and behaviour has improved. Despite being similar in nature, many achievement motivation approaches have been developed separately, suggesting that most achievement motivation theories are in concordance with one another instead of competing.

Motivational researchers have sought to promote a hierarchal model of approach and avoidance achievement motivation by incorporating the two prominent theories: the achievement motive approach and the achievement goal approach. Achievement motives include the need for achievement and the fear of failure. These are the more predominant motives that direct our behaviour towards positive and negative outcomes.

Achievement goals are viewed as more solid cognitive representations pointing individuals towards a specific end. There are three types of these achievement goals: a performance-approach goal, a performance-avoidance goal, and a mastery goal. A performance-approach goal is focused on attaining competence relative to others, a performance-avoidance goal is focused on avoiding incompetence relative to others, and a mastery goal is focused on the development of competence itself and of task mastery.

Achievement motives can be seen as direct predictors of achievement-relevant circumstances. Thus, achievement motives are said to have an indirect or distal influence, and achievement goals are said to have a direct or proximal influence on achievement-relevant outcomes. These motives and goals are viewed as working together to regulate achievement behaviour.

The hierarchal model presents achievement goals as predictors for performance outcomes. The model is being further conceptualized to include more approaches to achievement motivation. One weakness of the model is that it does not provide an account of the processes responsible for the link between achievement goals and performance. As this model is enhanced, it becomes more useful in predicting the outcomes of achievement-based behaviours.

Achievement Goals and Information Seeking

Theorists have proposed that people's achievement goals affect their achievement-related attitudes and behaviours. Two different types of achievement-related attitudes include task-involvement and ego-involvement. Task-involvement is a motivational state in which a person's main goal is to acquire skills and understanding whereas the main goal in ego-involvement is to demonstrate superior abilities.

One example of an activity where someone strives to attain mastery and demonstrate superior ability is schoolwork. However situational cues, such as the person's environment or surroundings, can affect the success of achieving a goal at any time. Studies confirm that a task-involvement activity more often results in challenging attributions and increasing effort (typically in activities providing an opportunity to learn and develop competence) than in an ego-involvement activity. Intrinsic motivation, which is defined as striving to engage in activity because of self-satisfaction, is more prevalent when a person is engaged in task-involved activities.

When people are more ego-involved, they tend to take on a different conception of their ability, where differences in ability limit the effectiveness of effort. Ego-involved individuals are driven to succeed by outperforming others, and their feelings of success depend on maintaining self-worth and avoiding failure. On the other hand, task-involved individuals tend to adopt their conception of ability as learning through applied effort. Therefore less able individuals will feel more successful as long as they can satisfy an effort to learn and improve. Ego-invoking conditions tend to produce less favourable responses to failure and difficulty.

Competence moderated attitudes and behaviours are more prevalent in ego-involved activities than task-involved. Achievement does not moderate intrinsic motivation in task-involving conditions, in which people of all levels of ability could learn to improve. In ego-involving conditions, intrinsic motivation was higher among higher achievers who demonstrated superior ability than in low achievers who could not demonstrate such ability. These different attitudes towards achievement can also be compared in information seeking. Task- and ego-involving settings bring about different goals, conceptions of ability, and responses to difficulty.

They also promote different patterns of information seeking. People of all levels of ability will seek information relevant to attaining their goal of improving mastery in task-involving conditions. However they need to seek information regarding self-appraisal to gain a better understanding of their self-capacity. On the other hand people in ego-involving settings are more interested in information about social comparisons, assessing their ability relative to others.

Self-Worth Theory in Achievement Motivation

Self-worth theory states that in certain situations students stand to gain by not trying and deliberately withholding effort. If poor performance is a threat to a person's sense of self-esteem, this lack of effort is likely to occur. This most often occurs after an experience of failure. Failure threatens self-estimates of ability and creates uncertainty about an individual's capability to perform well on a subsequent basis. If the following performance turns out to be poor, then doubts concerning ability are confirmed.

Self-worth theory states that one way to avoid threat to self-esteem is by withdrawing effort. Withdrawing effort allows failure to be attributed to lack of effort rather than low ability which reduces overall risk to the value of one's self-esteem. When poor performance is likely to reflect poor ability, a situation of high threat is created to the individual's intellect. On the other hand, if an excuse allows poor performance to be attributed to a factor unrelated to ability, the threat to self-esteem and one's intellect is much lower.

A study was conducted on students involving unsolvable problems to test some assumptions of the self-worth theory regarding motivation and effort. The results showed that there was no evidence of reported reduction of effort despite poorer performance when the tasks were described as moderately difficult as compared with tasks much higher in difficulty. The possibility was raised that low effort may not be responsible for the poor performance of students in situations which create threats to self-esteem. Two suggestions were made, one being that students might unconsciously withdraw effort, and the other stating that students may reduce effort as a result of withdrawing commitment from the problem. Regardless of which suggestion is true, self-worth theory assumes that individuals have a reduced tendency to take personal responsibility for failure.

Avoidance Achievement Motivation

In everyday life, individuals strive to be competent in their activities. In the past decade, many theorists have utilized a social-cognitive achievement goal approach in accounting for individuals striving for competence. An achievement goal is commonly defined as the purpose for engaging in a task, and the specific type of goal taken on creates a framework for how individuals experience their achievement pursuits.

Achievement goal theorists commonly identify two distinct ideas towards competence: a performance goal focused on demonstrating ability when compared to others, and a mastery goal focused on the development of competence and task mastery. Performance goals are hypothesized to produce vulnerability to certain response patterns in achievement settings such as preferences for easy tasks, withdrawal of effort in the face of failure, and decreased task enjoyment.

Mastery goals can lead to a motivational pattern that creates a preference for moderately challenging tasks, persistence in the face of failure, and increased enjoyment of tasks. Most achievement goal theorists conceptualize both performance and mastery goals as the "approach" forms of motivation. Existing classical achievement motivation theorists claimed that activities are emphasized and oriented towards attaining success or avoiding failure, while the achievement goal theorists focused on their approach aspect.

More recently, an integrated achievement goal conceptuali-zation was proposed that includes both modern performance and mastery theories with the standard approach and avoidance features. In this basis for motivation, the performance goal is separated into an independent approach component and avoidance component, and three achievement orientations are conceived: a mastery goal focused on the development of competence and task mastery, a performance-approach goal directed towards the attainment of favourable judegments of competence, and a performance-avoidance goal centered on avoiding unfavourable judegments of competence. The mastery and performance-approach goals are chara-cterized as self-regulating to promote potential positive outcomes and processes to absorb an individual in their task or to create excitement leading to a mastery pattern of achievement results. Performance-avoidance goals, however, are characterized as promoting negative circumstances. This avoidance orientation creates anxiety, task distraction, and a pattern of helpless achievement outcomes.

Intrinsic motivation, which is the enjoyment of and interest in an activity for its own sake, plays a role in achievement outcomes as well. Performance-avoidance goals undermined intrinsic motivation while both mastery and performance-approach goals helped to increase it. Most achievement theorists and philosophers also identify task-specific competence expectancies as an important variable in achievement settings.

Achievement goals are created in order to obtain competence and avoid failure. These goals are viewed as implicit (non-conscious) or self-attributed (conscious) and direct achievement behaviour. Competence expectancies were considered an important variable in classical achievement motivation theories, but now appear to only be moderately emphasized in contemporary perspectives.

Approach and Avoidance Goals

Achievement motivation theorists focus their research attention on behaviours involving competence. Individuals aspire to attain competence or may strive to avoid incompetence, based on the earlier approach-avoidance research and theories. The desire for success and the desire to avoid failure were identified as critical determinants of aspiration and behaviour by a theorist named Lewin.

In his achievement motivation theory, McClelland proposed that there are two kinds of achievement motivation, one oriented around avoiding failure and the other around the more positive goal of attaining success. Atkinson, another motivational theorist, drew from the work of Lewin and McClelland in forming his need-achievement theory, a mathematical framework that assigned the desire to succeed and the desire to avoid failure as important determinants in achievement behaviour.

Theorists introduced an achievement goal approach to achievement motivation more recently. These theorists defined achievement goals as the reason for activities related to competence. Initially, these theorists followed in the footsteps of Lewin, McClelland, and Atkinson by including the distinction between approach and avoidance motivation into the structure of their assumptions. Three types of achievement goals were created, two of which being approach orientations and the third an avoidance type. One approach type was a task involvement goal focused on the development of competence and task mastery, and the other being a performance or ego involvement goal directed towards attaining favourable judegments of competence. The avoidance orientation involved an ego or performance goal aimed at avoiding unfavourable judegments of competence. These new theories received little attention at first and some theorists bypassed them with little regard. Motivational theorists shifted away and devised other conceptualizations such as Dweck's performance-learning goal dichotomy with approach and avoidance components or Nicholls' ego and task orientations, which he characterized as two forms of approach motivation.

Presently, achievement goal theory is the predominant approach to the analysis of achievement motivation. Most contemporary theorists use the frameworks of Dweck's and Nicholls' revised models in two important ways. *First*, most theorists institute primary orientations towards competence, by either differentiating between mastery and ability goals or contrasting task and ego involvement.

A contention was raised towards the achievement goal frameworks on whether or not they are conceptually similar enough to justify a convergence of the mastery goal form (learning, task involvement and mastery) with the performance goal form (ability and performance, ego involvement, competition). *Secondly*, most modern theorists characterized both mastery and performance goals as approach forms of motivation, or they failed to consider approach and avoidance as independent motivational tendencies within the performance goal orientation.

The type of orientation adopted at the outset of an activity creates a context for how individuals interpret, evaluate, and act on information and experiences in an achievement setting. Adoption of a mastery goal is hypothesized to produce a mastery motivational pattern characterized by a preference for moderately challenging tasks, persistence in the face of failure, a positive stance towards learning, and enhanced task enjoyment.

A helpless motivational response, however, is the result of the adoption of a performance goal orientation. This includes a preference for easy or difficult tasks, effort withdrawal in the face of failure, shifting the blame of failure to lack of ability, and decreased enjoyment of tasks. Some theorists include the concept of perceived competence as an important agent in their assumptions. Mastery goals are expected to have a uniform effect across all levels of perceived competence, leading to a mastery pattern.

Performance goals can lead to mastery in individuals with a high perceived competence and a helpless motivational pattern in those with low competence. Three motivational goal theories have recently been proposed based on the tri-variant framework by achievement goal theorists: mastery, performance-approach, and performance-avoidance. Performance-approach and mastery goals both represent approach orientations according to potential positive outcomes, such as the attainment of competence and task mastery. These forms of behaviour and self-regulation commonly produce a variety of affective and perceptual-cognitive processes that facilitate optimal task engagement. They challenge sensitivity to information relevant to success and effective concentration in the activity, leading to the mastery set of motivational responses described by achievement goal theorists. The performance-avoidance goal is conceptualized as an avoidance orientation according to potential negative outcomes. This form of regulation evokes self-protective mental processes that interfere with optimal task engagement.

It creates sensitivity to failure-relevant information and invokes an anxiety-based preoccupation with the appearance of oneself rather than the concerns of the task, which can lead to the helpless set of motivational responses. The three goal theories presented are very process oriented in nature. Approach and avoidance goals are viewed as exerting their different effects on achievement behaviour by activating opposing sets of motivational processes.

Intrinsic Motivation and Achievement Goals

Intrinsic motivation is defined as the enjoyment of and interest in an activity for its own sake. Fundamentally viewed as an approach form of motivation, intrinsic motivation is identified as an important component of achievement goal theory.

Most achievement goal and intrinsic motivational theorists argue that mastery goals are facilitative of intrinsic motivation and related mental processes and performance goals create negative effects. Mastery goals are said to promote intrinsic motivation by fostering perceptions of challenge, encouraging task involvement, generating excitement, and supporting self-determination while performance goals are the opposite.

Performance goals are portrayed as undermining intrinsic motivation by instilling perceptions of threat, disrupting task involvement, and creating anxiety and pressure.

An alternative set of predictions may be derived from the approach-avoidance framework. Both performance-approach and mastery goals are focused on attaining competence and foster intrinsic motivation.

More specifically, in performance-approach or mastery orientations, individuals perceive the achievement setting as a challenge, and this likely will create excitement, encourage cognitive functioning, increase concentration and task absorption, and direct the person towards success and mastery of information which facilitates intrinsic motivation. The performance-avoidance goal is focused on avoiding incompetence, where individuals see the achievement setting as a threat and seek to escape it. This orientation is likely to elicit anxiety and withdrawal of effort and cognitive resources while disrupting concentration and motivation.

Personal Goals Analysis

In recent years, theorists have increasingly relied on various goal constructs to account for action in achievement settings. Four levels of goal representation have been introduced: task-specific guidelines for performance, such as performing a certain action, situation-specific orientations that represent the purpose of achievement activity, such as demonstrating competence relative to others in a situation, personal goals that symbolize achievement pursuits, such as getting good grades, and self-standards and future self-images, including planning for future goals and successes. These goal-based achievement motivation theories have focused almost exclusively on approach forms behaviour but in recent years have shifted more towards avoidance. Motivation is an important factor in everyday life. Our basic behaviours and feelings are affected by our inner drive to succeed over life's challenges while we set goals for ourselves. Our motivation also promotes our feelings of competence and self-worth as we achieve our goals.

It provides us with means to compete with others in order to better ourselves and to seek out new information to learn and absorb. Individuals experience motivation in different ways, whether it is task- or ego-based in nature. Some people strive to achieve their goals for personal satisfaction and self-improvement while others compete with their surroundings in achievement settings to simply be classified as the best. Motivation and the resulting behaviour are both affected by the many different models of achievement motivation.

These models, although separate, are very similar in nature and theory. The mastery and performance achievement settings each have a considerable effect on how an individual is motivated. Each theorist has made a contribution to the existing theories in today's achievement studies. More often than not, theorists build off of each other's work to expand old ideas and create new ones. Achievement motivation is an intriguing field, and I find myself more interested after reviewing similar theories from different perspectives.

POWER OF MOTIVATION

In the modern business environment, employers, team leaders, and managers are expected to understand and apply motivational techniques. Sales managers, in particular, must utilize powerful motivators to counteract the daily rejection experienced by members of the sales force. Some organizations use motivational tools to help their members meet the challenges of serving the public and conducting fundraising activities.

Fraternal organizations, Boy and Girl Scout troops and Church groups are examples of these. People who provide counselling services—Alcoholics Anonymous and Al-Anon, family and school counselors, group therapists, weight loss counselors, and self-help groups - rely heavily on motivational tools, such as goal setting, to keep clients on track, particularly in day-to-day situations. Athletic coaches use visualization and other techniques to help players improve and win.

BUILDING A REPERTOIRE

Although we know that different people respond to different motivators, a successful motivational speaker or seminar leader benefits most from having a broad repertoire of motivational tools and strategies.

TOOLS AND STRATEGIES

Goal Setting: People can be overwhelmed by goals that appear to be beyond their reach. Helping people set reasonable goals and break down long-term goals into shorter increments is helpful. The dieter who plans to lose 100 pounds may lack the motivation to diet for a year or more, but a loss of 2-3 pounds a week is an immediate goal that can be easily reached and reinforced with praise. Read more about Goal Setting on our helpful information site.

Focus: People who are successful in reaching goals are those who are able to maintain their focus, even when the objective is years away. Many of those successful people mention that they feel inspired to persist by focusing on a symbol of success. The symbol itself might be one that epitomizes success, uniqueness, or height (in our culture, reaching for success involves reaching up or skyward!).

Symbols of Success: The eagle is a great example of a lofty, free spirit that soars the rest and symbolizes power and success. Other people look for symbols that help them focus their energy. Many people consider pyramids as powerful, stable structures that help them focus their energies. Still others select photographs and other art forms that depict athletes or powerful animals moving towards their goals.

Music: A substantial research base exists to validate the power of music to motivate people. Many describe a favourite piece of music as "moving" or "inspiring." These descriptors indicate that music has great appeal to the emotional being, and that the effects of music penetrate beyond the conscious mind to the inner self where persistence and hope reside. Read more about the motivational power of music on this site.

SETTING GOALS

We've all been through it, time after time. We decide to make some changes in our lives: to lose weight, quit smoking, start exercising. We set our goals, and we work towards them. And, for a while, it works very well! We're dedicated, motivated, and moving towards improving ourselves. And then, slowly, perhaps so slowly we don't even notice it, we start to slip.

We skip a workout because we've booked an appointment for that time. We find excuses to light up just once, and that chocolate mountain fudge cheesecake just looks so good ... and before we know it, our careful plans are all for nothing, and we have to start again from scratch. Trouble is, every time we try again, we fail, and eventually the gap between starting and failing to reach a goal narrows, until finally, we just stop trying. Sound familiar? It happens to people all the time. It's not that people lack the self motivation or willpower to set goals and attain them. It's just that many of us have no idea how to set realistic goals, or how to maintain the motivation necessary to keep working towards them.

REACHING GOALS

Fortunately, if you watch successful people, you start to see some patterns—patterns anyone can apply. Successful people are masters at setting goals. They don't just announce, "I'm going to lose 20 lbs." Instead, they break the process down into easy-to-attain goals.

A twenty-pound loss may be their ultimate goal, but they start by determining to lose five. Losing five pounds sounds a lot easier than losing twenty! Once they've lost five, they set a new goal: to lose another five pounds, and so on until they meet their goal. There's an old saying that a journey of a thousand miles begins with a single step. It's often the thousand miles we tend to see, and not the individual steps towards the goal. Enjoy the journey towards your goals even more than actually attaining them, because it's the journey that ultimately makes setting goals worthwhile.

We all set goals, whether we're humming the tune to a favourite song, daydreaming of what could be, or making resolutions to create a significant change in our lives.

GOOD NEWS ABOUT GOAL SETTING

The good news is that we often attain our goals! Striving for more — greater success, more money, recognition, personal security, a better job — sometimes keeps up from acknowledging those goals we have reached. We're just too busy planning for the next phase.

Take a few minutes, as you read this, to stop and think about what you've achieved in your life. We're reminded, from time to time, of our greatest achievements simply because many of them are linked to concrete evidence: finishing a college degree, getting a large raise, giving birth to a child, visiting a foreign country. Perhaps at one time, these achievements were important goals, but they've gradually become a part of your life. Chances are, you've lost track of all the improvements you've made and aren't making a big deal of the goals you've reached.

AFFILIATION MOTIVATION

Affiliation motivation is the need for human relationships and for meaningful social contact. It was studied by David McClelland developed scales to measure it. Those with a high need for affiliation need harmonious relationships with other people and need to feel accepted by other people. They tend to conform to the norms of their work group.

High Affiliation motivation individuals prefer work that provides significant personal interaction. They enjoy being part of groups and when not anxious make excellent team members, though sometimes are distractible into social interaction. They can perform well in customer service and client interaction situations.

COMPETENCE MOTIVATION

Hull's motivational theory dominated the psychology of motivation during the late 1930s and the 1940s. Hull's last great work was published in 1952, shortly before he died. Already psychologists were questioning Hull's exclusive emphasis on drive-reduction. Hull portrayed all motivation as a matter of reducing uncomfortable drive or tension states. It is true that hunger and thirst are uncomfortable, but this conception of motivation leaves out many of the factors that energize and direct human behaviour.

As Hull's dominance in the field weakened, psychologists began to explore different types of motives not directly aimed at satisfying biological needs. In 1959 Robert W.

White wrote a classic substance for Psychological Review titled, "Motivation Reconsidered: The Concept of Competence." In it, White proposed a new concept: effectance motivation. Effectance was described as a "tendency to explore and influence the environment." White suggested that the "master reinforcer" for humans is personal competence.

He defined competence as "the ability to interact effectively with the environment." White pointed out several ways competence motives were different from Hullian motives. Unlike biological motives such as hunger and thirst, competence motives are never really satisfied.

They serve to enhance the abilities of the organism, rather than to regulate a biological process. They are not based on a state of biological deprivation. Rather, they help an organism improve itself. Competence motivation is visible in children. Toddlers try to act powerful and capable, big and grown up, almost as soon as they understand the concepts. Children of all ages try to exercise control over some domain of objects (whether it is a doll house, a collection of cars, or something else).

Healthy, normal children commonly wish to be regarded as knowledgeable and capable beyond their years. In general, people who have a special talent love to exercise it. People like a subject or a game that "plays to their strengths" because it makes them feel competent. Competency motivation emerges as a critical factor in career success. A survey of successful entrepreneurs—people who started their own businesses—showed two factors that were even more important than competitiveness or grades in school. The important factors were

(1) an appetite for hard work, and (2) an enjoyment of mastering skills. They were powerful predictors of financial success.

MASLOW'S HIERARCHY OF NEEDS

Psychologist Abraham Maslow first introduced his concept of a hierarchy of needs in his 1943 paper "A Theory of Human Motivation", Motivation and Personality. This hierarchy suggests that people are motivated to fulfill basic needs before moving on to other needs. Maslow's hierarchy of needs is most often displayed as a pyramid.

The lowest levels of the pyramid are made up of the most basic needs, while the more complex needs are located at the top of the pyramid. Needs at the bottom of the pyramid are basic physical requirements including the need for food, water, sleep and warmth. Once these lower-level needs have been met, people can move on to the next level of needs, which are for safety and security. As people progress up the pyramid, needs become increasingly psychological and social. Soon, the need for love, friendship and intimacy become important. Further up the pyramid, the need for personal esteem and feelings of accomplishment take priority. Like Carl Rogers, Maslow emphasized the importance of self-actualization, which is a process of growing and developing as a person to achieve individual potential.

TYPES OF NEEDS

Maslow believed that these needs are similar to instincts and play a major role in motivating behaviour. Physiological, security, social, and esteem needs are deficiency needs (also known as D-needs), meaning that these needs arise due to deprivation.

Satisfying these lower-level needs is important in order to avoid unpleasant feelings or consequences. Maslow termed the highest-level of the pyramid as growth needs (also known as being needs or B-needs). Growth needs do not stem from a lack of something, but rather from a desire to grow as a person.

FIVE LEVELS OF THE HIERARCHY OF NEEDS

There are five different levels in Maslow's hierarchy of needs:

1. *Physiological Needs*: These include the most basic needs that are vital to survival, such as the need for water, air, food and sleep. Maslow believed that these needs are the most basic and instinctive needs in the hierarchy because all needs become secondary until these physiological needs are met.
2. *Security Needs:* These include needs for safety and security. Security needs are important for survival, but they are not as demanding as the physiological needs. Examples of security needs include a desire for steady employment, health insurance, safe neighbourhoods and shelter from the environment.

3. *Social Needs:* These include needs for belonging, love and affection. Maslow considered these needs to be less basic than physiological and security needs. Relationships such as friendships, romantic attachments and families help fulfill this need for companionship and acceptance, as does involvement in social, community or religious groups.
4. *Esteem Needs:* After the first three needs have been satisfied, esteem needs becomes increasingly important. These include the need for things that reflect on self-esteem, personal worth, social recognition and accomplishment.
5. *Self-actualizing Needs:* This is the highest level of Maslow's hierarchy of needs. Self-actualizing people are self-aware, concerned with personal growth, less concerned with the opinions of others and interested fulfilling their potential.

WHAT IS SELF-ACTUALIZATION?

What exactly is self-actualization? Located at the peak of Maslow's hierarchy, he described this high-level need in the following way:

"What a man can be, he must be. This need we may call self-actualization...It refers to the desire for self-fulfillment, namely, to the tendency for him to become actualized in what he is potentially. This tendency might be phrased as the desire to become more and more what one is, to become everything that one is capable of becoming."

While Maslow's theory is generally portrayed as a fairly rigid hierarchy, Maslow noted that the order in which these needs are fulfilled does not always follow this standard progression. For example, he notes that for some individuals, the need for self-esteem is more important than the need for love. For others, the need for creative fulfillment may supersede even the most basic needs.

Characteristics of Self-Actualized People

In addition to describing what is meant by self-actualization in his theory, Maslow also identified some of the key characteristics of self-actualized people:

- *Acceptance and Realism*: Self-actualized people have realistic perceptions of themselves, others and the world around them.
- *Problem-centering*: Self-actualized individuals are concerned with solving problems outside of themselves, including helping others and finding solutions to problems in the external world. These people are often motivated by a sense of personal responsibility and ethics.
- *Spontaneity*: Self-actualized people are spontaneous in their internal thoughts and outward behaviour. While they can conform to rules and social expectations, they also tend to be open and unconventional.

- *Autonomy and Solitude*: Another characteristics of self-actualized people is the need for independence and privacy. While they enjoy the company of others, these individuals need time to focus on developing their own individual potential.
- *Continued Freshness of Appreciation*: Self-actualized people tend to view the world with a continual sense of appreciation, wonder and awe. Even simple experiences continue to be a source of inspiration and pleasure.
- *Peak Experiences*: Individuals who are self-actualized often have what Maslow termed peak experiences, or moments of intense joy, wonder, awe and ecstasy. After these experiences, people feel inspired, strengthened, renewed or transformed.

HERZBERG'S MOTIVATION-HYGIENE THEORY

To better understand employee attitudes and motivation, Frederick Herzberg performed studies to determine which factors in an employee's work environment caused satisfaction or dissatisfaction. He published his findings in the 1959 book *The Motivation to Work*.

The studies included interviews in which employees where asked what pleased and displeased them about their work. Herzberg found that the factors causing job satisfaction (and presumably motivation) were different from those causing job dissatisfaction. He developed the motivation-hygiene theory to explain these results. He called the satisfiers motivators and the dissatisfiers hygiene factors, using the term "hygiene" in the sense that they are considered maintenance factors that are necessary to avoid dissatisfaction but that by themselves do not provide satisfaction.

The following table presents the top six factors causing dissatisfaction and the top six factors causing satisfaction, listed in the order of higher to lower importance.

FACTORS AFFECTING JOB ATTITUDES

Leading to Dissatisfaction	Leading to Satisfaction
•Company policy	• Achievement
•Supervision	• Recognition
•Relationship with Boss	• Work itself
•Work conditions	• Responsibility
•Salary	• Advancement
•Relationship with Peers	• Growth

Herzberg reasoned that because the factors causing satisfaction are different from those causing dissatisfaction, the two feelings cannot simply be treated as opposites of one another. The opposite of satisfaction is not dissatisfaction, but rather, no satisfaction. Similarly, the opposite of dissatisfaction is no dissatisfaction.

While at first glance this distinction between the two opposites may sound like a play on words, Herzberg argued that there are two distinct human needs portrayed. *First*, there are physiological needs that can be fulfilled by money, for example, to purchase food and shelter. *Second*, there is the psychological need to achieve and grow, and this need is fulfilled by activities that cause one to grow.

From the table of results, one observes that the factors that determine whether there is dissatisfaction or no dissatisfaction are not part of the work itself, but rather, are external factors. Herzberg often referred to these hygiene factors as "KITA" factors, where KITA is an acronym for Kick In The A..., the process of providing incentives or a threat of punishment to cause someone to do something. Herzberg argues that these provide only short-run success because the motivator factors that determine whether there is satisfaction or no satisfaction are intrinsic to the job itself, and do not result from carrot and stick incentives.

IMPLICATIONS FOR MANAGEMENT

If the motivation-hygiene theory holds, management not only must provide hygiene factors to avoid employee dissatisfaction, but also must provide factors intrinsic to the work itself in order for employees to be satisfied with their jobs. Herzberg argued that job enrichment is required for intrinsic motivation, and that it is a continuous management process.

According to Herzberg:

- The job should have sufficient challenge to utilize the full ability of the employee.
- Employees who demonstrate increasing levels of ability should be given increasing levels of responsibility.
- If a job cannot be designed to use an employee's full abilities, then the firm should consider automating the task or replacing the employee with one who has a lower level of skill. If a person cannot be fully utilized, then there will be a motivation problem.

Critics of Herzberg's theory argue that the two-factor result is observed because it is natural for people to take credit for satisfaction and to blame dissatisfaction on external factors. Furthermore, job satisfaction does not necessarily imply a high level of motivation or productivity. Herzberg's theory has been broadly read and despite its weaknesses its enduring value is that it recognizes that true motivation comes from within a person and not from KITA factors.

ERG THEORY OF MOTIVATION

In 1969, Clayton Alderfer's revision of Abraham Maslow's Hierarchy of Needs, called the ERG Theory appeared in Psychological Review in an substance titled "An Empirical Test of a New Theory of Human Need."

Alderfer's contribution to organizational behaviour was dubbed the ERG theory (Existence, Relatedness, and Growth), and was created to align Maslow's motivation theory more closely with empirical research.

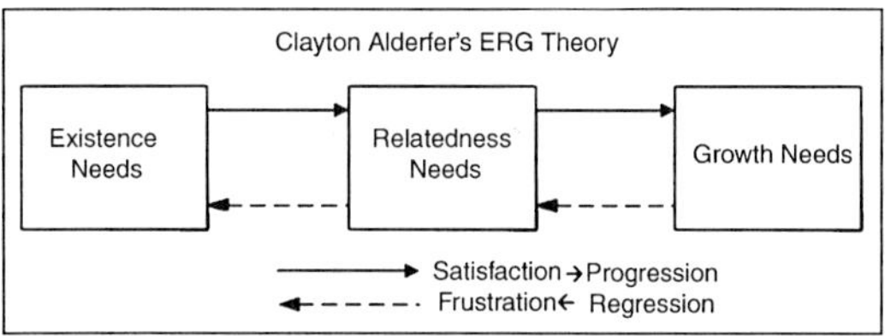

SIMILARITIES TO MASLOW'S NEEDS HIERARCHY

After the original formulation of Maslow's Hierarchy of Needs, studies had shown that the middle levels of Maslow's hierarchy overlap. Alderfer addressed this issue by reducing the number of levels to three.

The letters ERG represent these three levels of needs:

1. Existence refers to our concern with basic material existence motivators.
2. Relatedness refers to the motivation we have for maintaining interpersonal relationships.
3. Growth refers to an intrinsic desire for personal development.

Like Maslow's model, the ERG motivation is hierarchical, and creates a pyramid or triangle appearance. Existence needs motivate at a more fundamental level than relatedness needs, which, in turn supercedes growth needs.

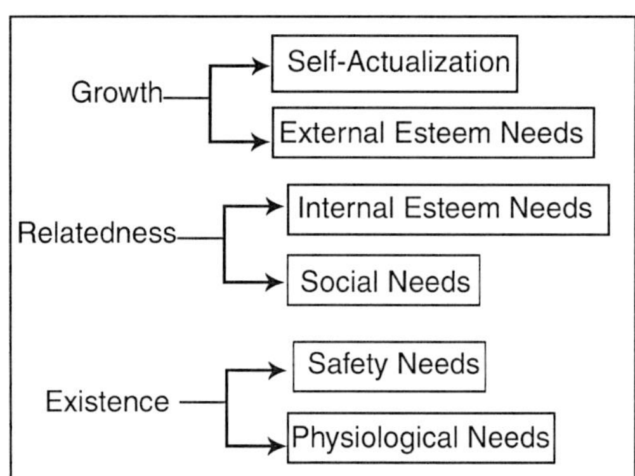

DIFFERENCES FROM MASLOW'S NEEDS HIERARCHY

Beyond simply reducing the distinction between overlapping needs, the ERG theory improves upon the following shortcomings of Maslow's Needs Hierarchy:

- Alderfers ERG theory demonstrates that more than one need may motivate at the same time. A lower motivator need not be substantially satisfied before one can move onto higher motivators.

- The ERG theory also accounts for differences in need preferences between cultures better than Maslow's Need Hierarchy; the order of needs can be different for different people. This flexibility accounts for a wider range of observed behaviours. For example, it can explain the "starving artist" who may place growth needs those of existence.
- The ERG theory acknowledges that if a higher-order need is frustrated, an individual may regress to increase the satisfaction of a lower-order need which appears easier to satisfy. This is known as the frustration-regression principle.

LEADERSHIP LESSONS

Unlike with Maslow's theory, managers need to understand that each employee operates with the need to satisfy several motivators simultaneously. Based upon the ERG theory, leadership which focuses on exlcusively one need at a time will not motivate their people effectively. Furthermore, the frustration-regression principle has additional impact on motivation in the workplace. As an example, if employees are not provided opportunities to grow, an employee might regress to fulfilling relatedness needs, socializing with co-workers more. Or, the inability of the environment or situation to satisfy a need for social interaction might increase the desire for more money or better working conditions. If Leadership recognizes these conditions soon enough in the process, they can take steps to satisfy those needs which are frustrated until such time that the worker can again pursue growth.

5

Implications of Industrial Planning

CONCEPT

Planning in organizations and public policy is both the organizational process of creating and maintaining a plan; and the psychological process of thinking about the activities required to create a desired goal on some scale. As such, it is a fundamental property of intelligent behaviour. This thought process is essential to the creation and refinement of a plan, or integration of it with other plans, that is, it combines forecasting of developments with the preparation of scenarios of how to react to them.

An important, albeit often ignored aspect of planning, is the relationship it holds with forecasting. Forecasting can be described as predicting what the future will look like, whereas planning predicts what the future should look like. The term is also used for describing the formal procedures used in such an endeavor, such as the creation of documents, diagrams, or meetings to discuss the important issues to be addressed, the objectives to be met, and the strategy to be followed. Beyond this, planning has a different meaning depending on the political or economic context in which it is used.

Two attitudes to planning need to be held in tension: on the one hand we need to be prepared for what may lie ahead, which may mean contingencies and flexible processes. On the other hand, our future is shaped by consequences of our own planning and actions.

ORGANISATIONS NEED A PLANNING ARCHITECTURE

A planning architecture is an overview of how different planning processes fit together.

It identifies:

- Different types of plan
- The time horizon of each
- When they have to be completed
- Time allowed for preparing the plan
- The frequency of updating

- Who is responsible
- How the different plans fit together.

Table. A Planning Architecture

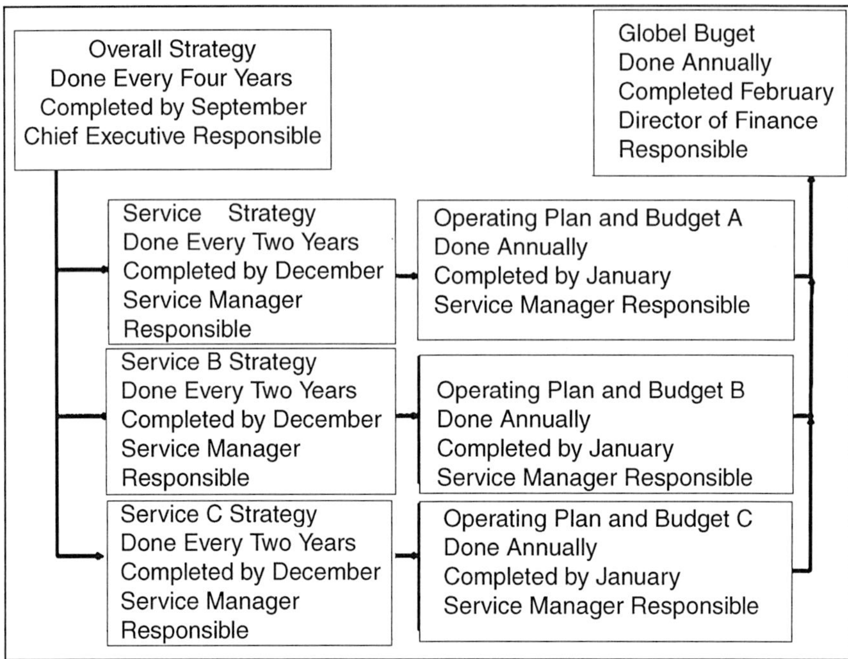

PLANNING IS AN INTELLECTUAL PROCESS

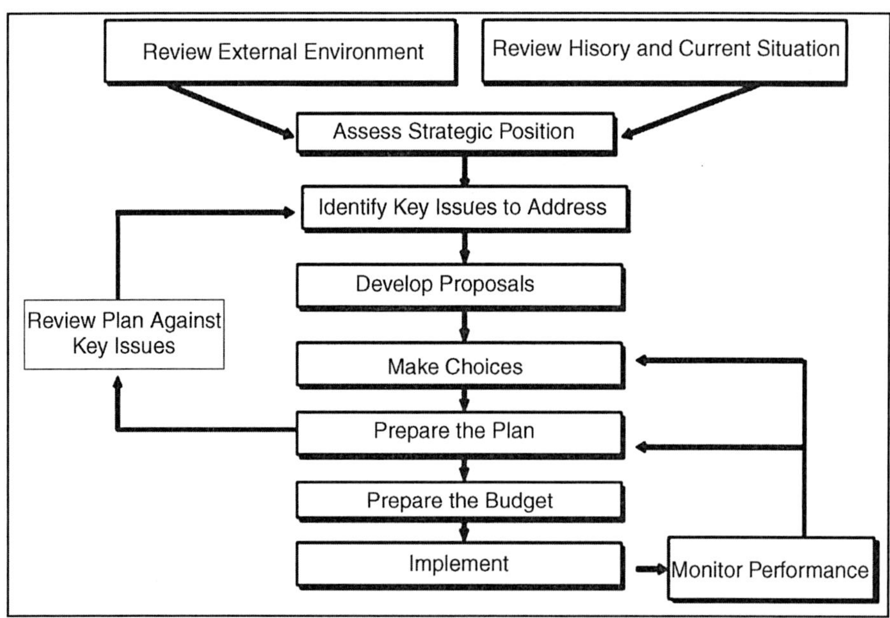

PLANNING IS A SOCIAL PROCESS

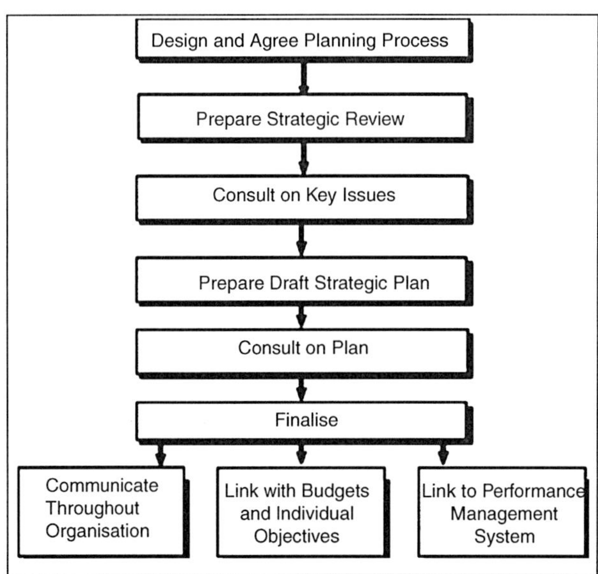

THE PLANNING PROCESS

Planning is one those things that we all know is good for us, but that no one wants to take the time to do. While it may seem that planning only takes time away from running your business, operating a business without a plan is like going to a grocery store without a list and trying to remember all the items that are needed. One comes out of the store having forgotten something critical - and having purchased a number of items that are totally frivolous and may never be used. It is the same for a business operating without a plan. Critical issues do not get addressed - and some tasks get done that have no relationship to the direction the business needs to go. For a business, however, the consequences of these unaddressed issues can range from inconvenience to bankruptcy.

Part of this reluctance is due to how complicated the process is viewed. Yet a complicated plan is almost as useless as none. The real question is how to make something simple that fits your business' needs. Can a good grocery list system be devised that isn't unnecessarily burdensome for all involved? Of course. Let's take a look as what planning really entails. The word "plan" originated from then Medieval Latin word planus which meant a level or flat surface. This evolved in French into being a map or a drawing of any object made by projection upon a flat surface. In English this has become a more general sense of a scheme of action, design or method. Planning in its current usage in business implies a consciousness of what is happening in the business. It does not preclude creativity or instinct, but it does add a layer of awareness that spells the difference between survival and extinction in a changing environment.

Planning does involve:

- An understanding of the business' history
- An examination of the business' environment
- An assessment of the business' mission
- Goals
- A process for reaching those goals
- A process for gathering information
- A realization that planning is a continuing process that is constantly evolving

Planning does not necessarily mean trying to project the future, but being aware of a range of likely futures and being prepared for them as occur.

BUSINESS PLAN

A business plan is used when one is starting a new business or a new process or product within a business. It includes not only a description of the new business, process or product, but also a discussion of how one plans on managing the marketing, development, production, and financing of this new venture.

ORGANIZATIONAL PLAN

Organizational planning, when it does occur, too often is spurred by crisis, focused on the short term, and not well thought out. To create healthy futures, organizations must construct processes for creating their futures that are not fueled by crisis and turmoil. It can be done.

One of the most confusing aspects for those who want to plan is the variety of terms that are used in conjunction with planning. How do you differentiate between a business plan, a financial plan, a marketing plan, a human resources plan, an operations plan, a strategic plan, a long-range plan and just plain general planning? The simple answer is that each area of your business needs planning so each area should have its own grocery list of what it wishes to accomplish in the future.

STRATEGIC PLAN

A strategic plan usually refers to the overall direction you wish your business to take over the longer term. Consequently, a long-range plan and a strategic plan are often used synonymously. Within that overall strategy a business will have shorter term financial goals, marketing goals, production goals, and human resource goals that will each need some type of plan if they are to be achieved. Just because a strategic plan is longer term does not mean it is never changed, however. One of the most serious mistakes businesses make is not revising their strategic plan regularly. The environment the business is operating in is changing constantly. The plan must be revisited at regular intervals to reflect the impact on the business of these external factors. There are some universal principles that are true across all types of planning. Before tackling more specific planning models, it is wise to gain an understanding of the basic principles of general planning.

PLANNING PRINCIPLES

Any plan should include who, what, when, where, how, and why:

- Who is needed to accomplish this task?
- What needs to be done?
- When does it start and end?
- Where will it take place?
- How will it happen?
- Why must we do it?

Along with the answers to these questions there needs to be some operational scheme to organize the tasks needed to achieve the goal. A helpful approach is to work backward from the goal to decide what must be done to reach it.

The backward approach is a way of looking at the big picture first, and then planning all tasks, conditions, and details in a logical sequence to make the big picture happen. From this a to-do list can easily be made. This list will become a checklist to ensure everything is progressing as planned. Adjustments can be made based on changing circumstances. The plan should be referenced often as a set of signposts on the journey towards the goals.

For many of us who left corporate America in favour of a smaller work environment, the idea of drafting a plan may seem offensive. After all, isn't frustration with all that busywork one of the reasons we left in the first place? We all have an aversion to doing anything on our job that doesn't immediately help the situation we are now experiencing.

However, isn't it also true that a little foresight and action before the fact can help eliminate many of the problems we face each day. Wouldn't it be nice to anticipate something like a price cut by your major competitor or a rise in the interest rate on your credit line? Of course it would. And with that anticipation comes an organized and effective response. That is what planning is really about.

TYPES OF PLANNING

CORPORATE PLANNING

From a company's perspective, corporate planning involves formulating long term business goals so that the strategic planning of an enterprise may be developed and acted upon. The corporate planning term that was popular in the 1960s has since been referred to as strategic management.

Primary Focus

Corporate and strategic planning focuses on efficiently and effectively achieving optimal performance from all business divisions, with actionable information provided to executive management in order to foster accountability and business growth.

Process

The process of gathering data to prepare detailed action plans addresses a corporate organization's goals and objectives so that an actionable plan may be developed, usually by senior management.

Information Considered

The resources of a business, company or organization is considered along with the environment in which it is operating. Statistical data including budget formulation and performance management is used in connection with business analysis and forecasting.

Planning

A plan is developed using the determined information and resources including targets and milestones that corporate planning executives and managers may use to reach company goals and objectives.

Significance

As a formal and structured approach, corporate planning can be instrumental in achieving strategic corporate goals and objectives for any kind of business organization.

FUNCTIONAL AND CROSS-FUNCTIONAL PLANNING

Functional Planning

In order to understand how best to manage functional planning activities, we need to understand what they are and what their primary objectives are. We also need to understand how they differ from cross-functional plans. A functional plan essentially manages the work of a department that is concerned with performing the same type of work for all projects as well as performing non-project work.

For project related work, the functional plan:

- *Defines tasks at the level they can be assigned to an individual*: Individuals in a functional organization must have a clear and unambiguous understanding of their work assignments.
- *Produces clearly defined final outputs to other functional organizations*: Clearly defined outputs helps ensure that needs of the programme are satisfied, and the functional organization has a consistent

checkpoint for measuring throughput of deliverables across the entire portfolio of programmes.

- *Tracks internal deliverables*: Outputs that are never delivered to another functional activity or reported at the programme level may nevertheless be important in order to move work within an organization.
- *Includes a process for incorporating programme requirements*: A functional plan must "understand" when deliverables are required by a programme, which deliverables are shared by multiple programmes, and which programme requirements are most restrictive at any particular time.
- *Includes a process for managing competing priorities from multiple programmes*: Functional resources may often be strained to capacity when many programmes compete for attention. Functional planning must incorporate processes that ensure the right programme gets the right work done at the right time, and each programme must also understand its position in the priority.
- *May be impacted by non-project related work*: Functional planning may be impacted by internal departmental activity such as training, process improvement initiatives, or other internal "projects".
- *Includes change work*: Functional plans must also devote resources to managing changes to existing work products. For example, product designs are often revised as a result of new cost initiatives, test results or redesign of nearby components.
- *Includes a process for assessing plan status*: Functional planners must have clearly defined processes for statusing work as complete. This is particularly important in the case of external deliverables.

In short, functional planning supports the work, costs, and resources of a particular department.

This plan is often managed within the context of an overall programme plan, but must also consider competing priorities from multiple programmes, as well as internal initiatives and priorities such as process improvements and training requirements.

Cross-Functional Planning

Cross-functional planning brings together the activities of various functional groups in support of a single project. In the automotive industry, these projects are product development programmes, whether the product is an entire vehicle or a sub-system or component. Whereas functional planning focuses on moving work through the organization, cross-functional planning focuses on moving work from functional organization to functional organization. Given this objective, cross-functional planning will focus on managing the external outputs of a function - the key hand-offs between functions.

Cross-functional planning can be summarized as follows:

- *Work carried out by a single functional organization can be represented as a single activity*: The internal deliverables of a functional plan or schedule are not appropriate within the cross-functional plan. They are not the primary focus of the plan, which is to move work between departments. In cases where multiple deliverables are produced, additional activities may be included. For example, three major iterations of a product design may each be tracked, even though the work is performed by a single organization.
- *Completion of an activity represents a hand-off to another organization*: This hand-off is required for the downstream functional group to either begin work, continue work or complete work. For example, the product design is required to begin developing the part tools.
- *Includes a process for establishing programme requirements*: The requirements passed down to a functional plan must meet the needs of the programme, but it must also represent any initial trade-offs made among different functional groups. This allows the functional groups to share the risk inherent in any programme plan by distributing any available float time equitably among the deliverable-producing functional group and the deliverable-using functional group.
- *Focuses on planned activity*: Whereas the functional plan considers change work, the cross-functional focuses on planned work. Taking the example of a product design, the cross-functional plan would track the initial product design only and ignore the ongoing revisions. The timing impact of those revisions will of course manifest themselves in longer tooling times.
- *Includes a process for assessing the impact of plan status*: Understanding the status of activity progress is key to both functional and crossfunctional planning. For cross-functional plan, however, the focus is on understanding the impact to downstream activities in order, for example, to develop recovery plans. Where practical, cross-function plans should interface to functional schedules to obtain activity status.

Table. Functional and Cross Functional Planning Comparison

Cross-functional	Functional
Project oriented	Department oriented
Manages deliverables	Manages activities
Focus on requirements and risk	Focus on throughput and resources
Uses critical path scheduling database techniques	Uses date management and
Focuses on planned activity	Includes change work
Focus on moving work between departments	Focus on moving work through a department

Levels of Planning

Cross-functional and functional planning activities can be broadly categorized as occurring at two levels–programme and component. At the programme level, plan activities represent a summary. At the component level, activities are expanded out against the product scope.

For example, production tooling may be represented as a single activity at the programme level, but will appear as a separate activity for each production tool at the component level. Vehicle level activities that are not performed against the product scope will not appear in the component plans, other than as reference points. Interactions between the plan types are shown in Figure.

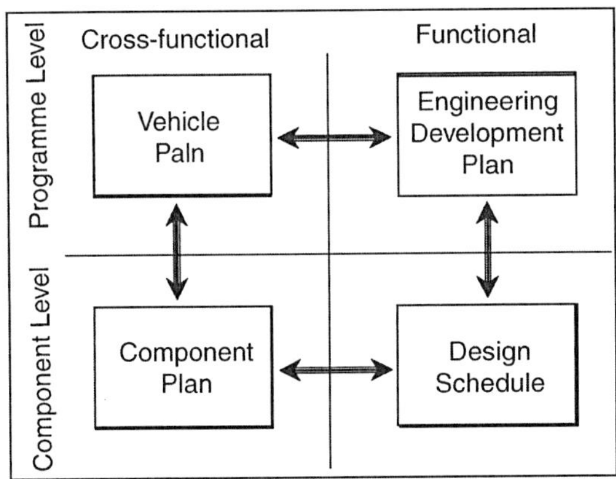

Fig. Level of Planning

Functional Plans Versus Functional Schedules

The authors make a distinction between different types of functional planning activities. This is particularly important in a product development project executed by a matrixed organization. Some clarification is required. A functional plan describes the activity of a given functional area for a given project.

For example, if the overall project is to develop and launch a new vehicle, then the functional plans might include:

- Purchasing/Sourcing Plan
- Engineering Development Plan
- Manufacturing Launch Plan
- Marketing/Sales Plan
- Quality Management Plan

Activities in a functional plan are summarized to the programme level. In addition to describing how a function will support the overall programme, functional plans also serve as a framework for functional schedules. Functional plans may also be created for internal department projects, not related to a specific vehicle programme. A functional schedule represents the work of a given functional area across many programmes. The summarized activities in the functional plan are broken down to discrete work elements, allowing for resource management. Functional schedules are used to manage workload, enable functional metrics, and support functional and other plan types.

In our example of a vehicle programme, functional schedules may include:

- Surface Release Schedule
- Sourcing Schedules
- Design Schedules
- Tooling Schedules
- Test Property Build Schedules

Managing the Relationship of the Programme Plan to Functional Plans

The programme level plan includes those events that are required to provide the organization with a clear understanding of the programme scope, timing and deliverables. It includes key events such as the launch date, and transition dates between project phases. It must, however, also serve to manage interfaces between functional plans.

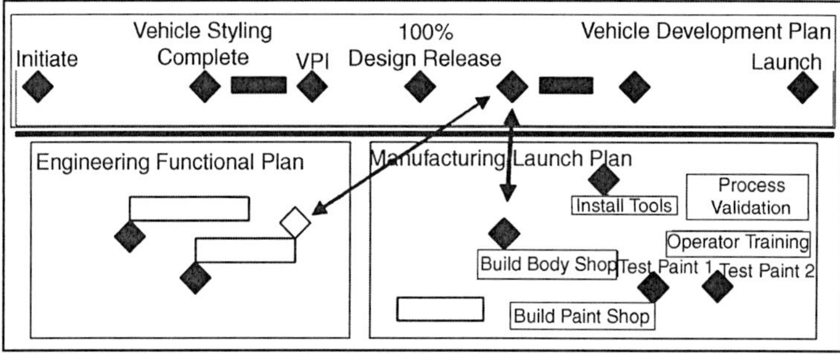

Fig. Managing Functional Plan Interfaces

For example, Figure describes a simplified relationship between the engineering functional plan and the manufacturing launch plan. The engineering plan contains product and process development activities, including the development of the assembly process and workflow, materials handling, and even workspace allocation.

All of these are critical to developing a successful body assembly facility. The vehicle programme plan does not need to track the functional activities, but rather it needs to manage the interfaces between them.

The requirement to begin building or retrofitting an assembly facility will dictate a requirement date for the development of a manufacturing process. In order to ensure that the interface is realistic, it is important that functional plans and the vehicle programme plan be developed together. Both the programme teams and the functional representatives should be well aware of their interface points and should focus on negotiating realistic hand-off dates.

These dates are then reflected in the vehicle programme plan and both functional plans are measured to this agreement. Managing the interface of the two functional plans through the cross-functional vehicle plan ensures that both satisfy the programme requirements.

Managing the Relationship of Functional Plans to Functional Schedules

Much as the overall programme plan is used to direct the development of component level crossfunctional plans, the functional plans serve as the basis for development of functional schedules within the context of the overall programme plans.

The functional plan lays out the department's strategy for the particular programme. Any new initiatives, such as the use of a new Computer Aided Design tool, are comprehended. Resource allocation is laid out at a high level, and blocks of time are established for the various departmental activities.

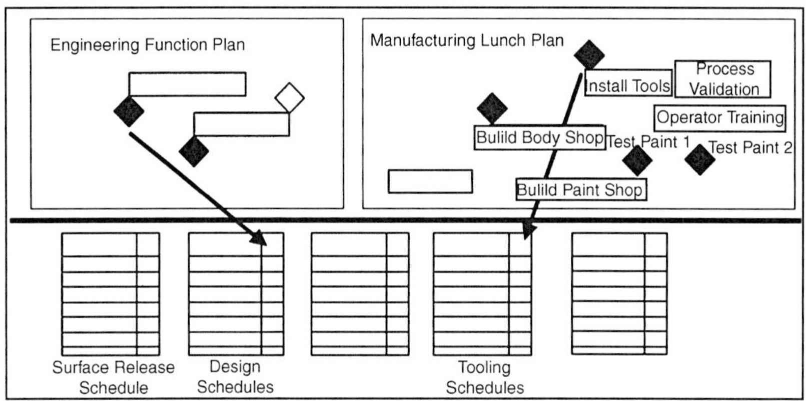

Fig. Managing Functional Schedules

The functional plan directs the development of the functional schedule for the programme. Any adjustments in work processes must be comprehended in the functional schedule in a way that allows the data to viewed and consolidated across programmes. Summarized activities are expanded to represent the detail scope of work, and resource allocations are translated into specific resource assignments.

Managing the Relationship of Cross-functional Component Plans to Functional Schedules

Component level plans managed by Product Development Teams at automotive OEMs, often called "PDT Plans", represent the plan to engineer a particular subsystem of an overall vehicle. The effort involves coordinating requirement development, styling, design, process development, validation, and manufacturing activities. The functional plan directs the development of functional schedules in terms of department strategy, and by ensuring integration with the cross-functional vehicle plan.

The PDT Plan serves a similar purpose in ensuring that the functional schedule supports the cross-functional development plan for the component and subsystem. The PDT Plan performs different roles during the planning verses execution phases of a programme. During the planning stages it is used to integrate product based functional schedules by refining requirements by commodity and by providing an end-to-end view of sub-system and component development. The end result is an integrated plan by commodity. Once execution begins, the PDT Plan is used to assess functional schedule performance to plan, identify potential risk areas, and to renegotiate agreements as needed. Activity tracking and forecasting is performed within the functional schedules. CPM tools and techniques are not as important but are still useful for projecting downstream impacts to the baseline plan.

STRATEGIC PLANNING

Strategic planning is an organization's process of defining its strategy, or direction, and making decisions on allocating its resources to pursue this strategy, including its capital and people. Various business analysis techniques can be used in strategic planning, including SWOT analysis, GE/McKinsey portfolio analysis, COPE analysis, PEST analysis, STEER analysis and EPISTEL. Strategic planning is the formal consideration of an organization's future course.

All strategic planning deals with at least one of three key questions:

1. "What do we do?"
2. "For whom do we do it?"
3. "How do we excel?"

In business strategic planning, some authors phrase the third question as "How can we beat or avoid competition?" But this approach is more about defeating competitors than about excelling. In many organizations, this is viewed as a process for determining where an organization is going over the next year or—more typically—3 to 5 years, although some extend their vision to 20 years. In order to determine where it is going, the organization needs to know exactly where it stands, then determine where it wants to go and how it will get there. The resulting document is called the "strategic plan." While strategic planning may be used to effectively plot a company's longer-term direction, one cannot use it to reliably forecast how the market will evolve and what issues will surface in the immediate future. Therefore, strategic innovation and tinkering with the "strategic plan" have to be a cornerstone strategy for an organization to survive the turbulent business climate.

Vision Statements, Mission Statements and Values

- *Vision*: Defines the way an organization or enterprise will look in the future. Vision is a long-term view, sometimes describing how the organization would like the world to be in which it operates. For example, a charity working with the poor might have a vision statement which reads "A World without Poverty."

- *Mission*: Defines the fundamental purpose of an organization or an enterprise, succinctly describing why it exists and what it does to achieve its Vision. It is sometimes used to set out a "picture" of the organization in the future. A mission statement provides details of what is done and answers the question: "What do we do?" For example, the charity might provide "job training for the homeless and unemployed."
- *Values*: Beliefs that are shared among the stakeholders of an organization. Values drive an organization's culture and priorities and provide a framework in which decisions are made. For example, "Knowledge and skills are the keys to success" or "give a man bread and feed him for a day, but teach him to farm and feed him for life". These example values may set the priorities of self sufficiency over shelter.
- *Strategy*: Strategy, narrowly defined, means "the art of the general". A combination of the ends for which the firm is striving and the means by which it is seeking to get there. A strategy is sometimes called a roadmap which is the path chosen to plow towards the end vision. The most important part of implementing the strategy is ensuring the company is going in the right direction which is towards the end vision.

Organizations sometimes summarize goals and objectives into a mission statement and/or a vision statement. Others begin with a vision and mission and use them to formulate goals and objectives. While the existence of a shared mission is extremely useful, many strategy specialists question the requirement for a written mission statement. However, there are many models of strategic planning that start with mission statements, so it is useful to examine them here.

- A Mission statement tells you the fundamental purpose of the organization. It defines the customer and the critical processes. It informs you of the desired level of performance.
- A Vision statement outlines what the organization wants to be, or how it wants the world in which it operates to be. It concentrates on the future. It is a source of inspiration. It provides clear decision-making criteria.

An advantage of having a statement is that it creates value for those who get exposed to the statement, and those prospects are managers, employees and sometimes even customers. Statements create a sense of direction and opportunity. They both are an essential part of the strategy-making process. Many people mistake the vision statement for the mission statement, and sometimes one is simply used as a longer term version of the other. The Vision should describe why it is important to achieve the Mission. A Vision statement defines the purpose or broader goal for being in existence or in the business and can remain the same for decades if crafted well.

A Mission statement is more specific to what the enterprise can achieve itself. Vision should describe what will be achieved in the wider sphere if the organization and others are successful in achieving their individual missions. A mission statement can resemble a vision statement in a few companies, but that can be a grave mistake. It can confuse people. The mission statement can galvanize the people to achieve defined objectives, even if they are stretch objectives, provided it can be elucidated in SMART terms.

A mission statement provides a path to realise the vision in line with its values. These statements have a direct bearing on the bottom line and success of the organization. Which comes first? The mission statement or the vision statement? That depends. If you have a new start up business, new programme or plan to re-engineer your current services, then the vision will guide the mission statement and the rest of the strategic plan.

If you have an established business where the mission is established, then many times, the mission guides the vision statement and the rest of the strategic plan. Either way, you need to know your fundamental purpose - the mission, your current situation in terms of internal resources and capabilities and external conditions, and where you want to go - the vision for the future. It's important that you keep the end or desired result in sight from the start.

Features of an effective vision statement include:

- Clarity and lack of ambiguity
- Vivid and clear picture
- Description of a bright future
- Memorable and engaging wording
- Realistic aspirations
- Alignment with organizational values and culture

To become really effective, an organizational vision statement must become assimilated into the organization's culture. Leaders have the responsibility of communicating the vision regularly, creating narratives that illustrate the vision, acting as role-models by embodying the vision, creating short-term objectives compatible with the vision, and encouraging others to craft their own personal vision compatible with the organization's overall vision. In addition, mission statements need to be subjected to an internal assessment and an external assessment.

The internal assessment should focus on how members inside the organization interpret their mission statement. The external assessment—which includes all of the businesses stakeholders—is valuable since it offers a different perspective. These discrepancies between these two assessments can give insight on the organization's mission statement effectiveness. Another approach to defining Vision and Mission is to pose two questions.

Firstly, "What aspirations does the organization have for the world in which it operates and has some influence over?", and following on from this, "What can the organization do or contribute to fulfill those aspirations?". The succinct answer to the first question provides the basis of the Vision Statement. The answer to the second question determines the Mission Statement.

Strategic Planning Outline

The preparatory phase of a business plan relies on planning. The first stages of a business plan include Analysis of the Current Situation and Marketing Plan Strategy and Objectives.

Analysis of the current situation - past year:

- Business trends analysis
- Market analysis
- Competitive analysis
- Market segmentation
- Marketing-mix
- SWOT analysis
- Positioning - analyzing perceptions
- Sources of information

Marketing plan strategy and objectives - next year

- Marketing strategy
- Desired market segmentation
- Desired marketing-mix
- TOWS-based objectives as a result of the SWOT
- Position and perceptual gaps
- Yearly sales forecast

According to Arieu, "there is strategic consistency when the actions of an organisation are consistent with the expectations of management, and these in turn are with the market and the context"

Strategic Planning Process

There are many approaches to strategic planning but typically a three-step process may be used:

1. *Situation*: Evaluate the current situation and how it came about.
2. *Target*: Define goals and/or objectives
3. *Path/Proposal*: Map a possible route to the goals/objectives

One alternative approach is called Draw-See-Think

- *Draw*: What is the ideal image or the desired end state?
- *See*: What is today's situation? What is the gap from ideal and why?
- *Think*: What specific actions must be taken to close the gap between today's situation and the ideal state?
- *Plan*: What resources are required to execute the activities?

An alternative to the Draw-See-Think approach is called See-Think-Draw

- *See*: What is today's situation?
- *Think*: Define goals/objectives
- *Draw*: Map a route to achieving the goals/objectives

Strategic Planning Tools and Approaches

Among most useful tools for strategic planning is SWOT analysis. The main objective of this tool is to analyse internal strategic factors, strengths and weaknesses attributed to the organization, and external factors beyond control of the organization such as opportunities and threats. Besides SWOT analysis, portfolio analyses such as the GE/McKinsey matrix or COPE analysis. can be performed to determine the strategic focus. Other tools include Balanced Scorecards, which creates a systematic framework for strategic planning, and scenario planning, which was originally used in the military and recently used by large corporations to analyse future scenarios.

Situational Analysis

When developing strategies, analysis of the organization and its environment as it is at the moment and how it may develop in the future, is important. The analysis has to be executed at an internal level as well as an external level to identify all opportunities and threats of the external environment as well as the strengths and weaknesses of the organizations.

There are several factors to assess in the external situation analysis:

- Markets
- Competition
- Technology
- Supplier markets
- Labour markets
- The economy
- The regulatory environment

It is rare to find all seven of these factors having critical importance. It is also uncommon to find that the first two - markets and competition - are not of critical importance. Analysis of the external environment normally focuses on the customer. Management should be visionary in formulating customer strategy, and should do so by thinking about market environment shifts, how these could impact customer sets, and whether those customer sets are the ones the company wishes to serve. Analysis of the competitive environment is also performed, many times based on the framework suggested by Michael Porter.

Goals, Objectives and Targets

Strategic planning is a very important business activity. It is also important in the public sector areas such as education. It is practiced widely informally and formally. Strategic planning and decision processes should end with objectives and a roadmap of ways to achieve them. One of the core goals when drafting a strategic plan is to develop it in a way that is easily translatable into action plans. Most strategic plans address high level initiatives and over-arching goals, but don't get articulated into day-to-day projects and tasks that will be required to achieve the plan. Terminology or word choice, as well as the level a plan is written, are both examples of easy ways to fail at translating your strategic plan in a way that makes sense and is executable to others. Often, plans are filled with conceptual terms which don't tie into day-to-day realities for the staff expected to carry out the plan. The following terms have been used in strategic planning: desired end states, plans, policies, goals, objectives, strategies, tactics and actions. Definitions vary, overlap and fail to achieve clarity.

The most common of these concepts are specific, time bound statements of intended future results and general and continuing statements of intended future results, which most models refer to as either goals or objectives. One model of organizing objectives uses hierarchies. The items may be organized in a hierarchy of means and ends and numbered as follows: Top Rank Objective, Second Rank Objective, Third Rank Objective, etc. From any rank, the objective in a lower rank answers to the question "How?" and the objective in a higher rank answers to the question "Why?" The exception is the Top Rank Objective: there is no answer to the "Why?" question. That is how the TRO is defined. People typically have several goals at the same time. "Goal congruency" refers to how well the goals combine with each other. Does goal A appear compatible with goal B? Do they fit together to form a unified strategy?

"Goal hierarchy" consists of the nesting of one or more goals within other goal. One approach recommends having short-term goals, medium-term goals, and long-term goals. In this model, one can expect to attain short-term goals fairly easily: they stand just slightly above one's reach. At the other extreme, long-term goals appear very difficult, almost impossible to attain. Strategic management jargon sometimes refers to "Big Hairy Audacious Goals" in this context. Using one goal as a stepping-stone to the next involves goal sequencing. A person or group starts by attaining the easy short-term goals, then steps up to the medium-term, then to the long-term goals. Goal sequencing can create a "goal stairway". In an organizational setting, the organization may co-ordinate goals so that they do not conflict with each other. The goals of one part of the organization should mesh compatibly with those of other parts of the organization.

OPERATIONAL PLANNING

The Operational Plan is the third part of your completed Strategic Plan. It defines how you will operate in practice to implement your action and monitoring plans–what your capacity needs are, how you will engage resources, how you will deal with risks, and how you will ensure sustainability of the project's achievements Operational planning is the process of assuring that specific tasks are carried out effectively. An operational planning is a subset of strategic work plan. An operational plan is the basis for, and justification of an annual operating budget request. Therefore, a five-year strategic plan would need five operational plans funded by five operating budgets.

Like a strategic plan, an operational plan addresses four questions:

1. Where are we Stand now ?
2. What do we want to Achieve?
3. How do we get there?
4. How do we measure our progress?

Operational plans should be prepared by the people who will be involved in implementation. There is often a need for significant cross-departmental dialogue as plans created by one part of the organisation inevitably have implications for other parts.

A operational planning allow us:

- Achievement of Goals
- Concentration of effort
- Clarity around cost/expenditure issues
- Confidence of staff
- Enhanced partnership working
- Assurance of success for Health
- Improvement/Development for Leads/Managers
- Clarity around cost/expenditure issues

Key Components of Operational Planning

- *Human Requirements:* The human capacity and skills required to implement any project, and your current and potential sources of these resources. A good and quality human resource give power to achieve any goal within the timelimit.
- *Financial Requirements*: finance is always needed for any planning and project. With operational planning we able to find the way of manage finance resource
- *Risk Assessment*: What risks exist and how they can be addressed. calculate your all the risk available.
- *Exit Strategy*: In exit strategy you plan when and how you will exit your project

Guidelines for Operational Planning

Following are the Guidelines for Operational Planning:

- What is the optimal operational plan to meet the specific system objectives, consistant with some long term plan, with existing facilities in the next planning period.

- What is the best operating plan on which, to base plan for production
- What specific operations or sequences of operations should be performed with existing facilities or meet specific output requirement in the next operational period.

LONG TERM PLANNING

Long Term Planning in Business is one of the most challenging tasks faced by the modern business executive. Long Term Planning is usually for a time horizon greater than three years and deals with strategic planning. The main difficulty with this type of business planning is trying to forecaste what will be happening in the market over extended periods. Some companies merely extrapolate their historical results, play around with the numbers a bit and believe they have produced a long term plan.

At the other end of the scale sophisticated forecasting software, scenario planning and PEST analysis is used. The mistake a lot of companies make is treating the long term plan as a chore that has to be completed periodically but that can then be ignored. A long term plan needs to be a living process and source of reference if it is to be effective. It has to be kept up to date reflecting changes in the PEST analysis and taking into account competitors strategies and tactics as they become known. The real value of long term planning is not in the plan itself, which will need to be changed, but in the process of planning, thinking ahead and considering the potential impact of outside influences on your business.

PLANNING BY DIRECTION AND PLANNING BY INDUCEMENT

Planning by direction is an integral part of a socialist society. It entails complete absence of laissez-faire. There is one central authority which plans, directs, and orders the execution of the plan in accordance with pre-determined targets and priorities. Such planning is comprehensive and encompasses the entire economy. Planning by inducement is democratic planning.

It means planning by manipulating the market. There is no compulsion but persuasion. There is freedom of enterprise, freedom of consumption and freedom of production. But these freedoms are subject to state control and regulation. People are induced to act in a certain way through various monetary and fiscal measures. Thus, planning by inducement is able to achieve the same results as are likely to be achieved in planning by direction but with less sacrifice of individual liberty.

FINANCIAL PLANNING AND PHYSICAL PLANNING

Financial planning refers to the technique of planning in which resources are allocated in terms of money. Financial planning is essential in order to remove maladjustments between supplies and demand and for calculating costs and benefits of the various projects.

Thus, Financial planning is thought to secure a balance between demands and supplies, avoid inflation and bring about economic stability. Physical planning refers to the allocation of resources in terms of men, materials and machinery. In physical planning, an overall assessment is made of the available real resources such as raw materials, manpower, etc., and how they have to be obtained so that bottlenecks may be eliminated during the plan. Physical planning requires the fixation of physical targets with regard to agricultural and industrial production, socio-cultural and transportation services, consumption levels and in respect of employment, income and investment levels of the economy. Physical planning has to be viewed as an overall long-term planning rather than a short-term piecemeal planning.

PERSPECTIVE PLANNING AND ANNUAL PLANNING

Perspective planning refers to long-term planning in which long range targets are set in advance for a period of 15, 20, or 25 years. A perspective plan, however, does not imply one plan for the entire period of 15 or 20 years. In reality, the broader objectives and targets are to be achieved within the specified period of time by dividing the perspective plan into several short-period plans of 4, 5 or 6 years.

Not only this, a five year plan is further broken up into annual plans so that each annual plan fits into the broad framework of the five-year plan. Plans of either kind are further divided into regional and sectoral plans. Regional plans pertain to regions, districts and localities and sectoral plans pertain to plans for agriculture, industry, foreign trade etc.

INDICATIVE PLANNING AND IMPERATIVE PLANNING

This is the French system of planning which is based on the principle of decentralization in the operation and execution of the national plans. This type of planning is not imperative but flexible. In indicative planning the private sector is neither rigidly controlled nor directed to fulfill the targets and priorities of the plan. Even then, the private sector is expected to fulfill the targets for the success of the plan.

The state provides all types of facilities to the private sector but does not direct it, rather indicates the areas in which it can help in implementing the plan. On the other hand, under imperative planning all economic activities and resources of the economy operate under the direction of the state. There is complete control over the factors of production by the state. The entire resources of the country are used to the maximum in order to fulfill the targets of the plan. There is no consumers' sovereignty in such planning. What and how much to produce–such decisions are taken by the managers of firms and factories on the direction of the planning commission or a central planning authority. Since the government policies and decisions are rigid, they cannot be changed easily.

DEMOCRATIC PLANNING AND TOTALITARIAN PLANNING

In totalitarian or authoritarian planning there is central control and direction of all economic activity in accordance with a single plan. There is planning by direction where consumption, production, exchange, and distribution are all controlled by the state. In totalitarian planning, the planning authority is the supreme body. It decides about the targets, schemes, allocations, methods and procedures of implementation of the plan.

There is absolutely no opposition to the plan. People have to accept and rigidly implement the plan. In democratic planning, the philosophy of democratic government is accepted as the ideological basis. People are associated at every step in the formulation and implementation of the plan. Cooperation of different agencies, and voluntary groups, and associations plays a major role in the execution of the plan. Democratic planning respects the institution of private property. Price mechanism is allowed to play its due role.

The government only seeks to influence economic and investment decisions in the private sector through fiscal and monetary measures. The private sector operates side by side with the public sector. Democratic planning aims at the removal of inequalities of income and wealth through peaceful means by taxation and government spending on social welfare and social security schemes. Individual freedom prevails and people enjoy social, economic and political freedoms.

ROLLING AND FIXED PLANS

In a rolling plan, every year three new plans are made and acted upon. First, there is a plan for the current year which includes the annual budget and the foreign exchange budget. Second, there is a plan for a number of years, say three, four or five. Third, a perspective plan for 10, 15 or 20 or even more years is presented every year in which the broader goals are stated and the outlines of future development are forecast. The annual one-year plan is fitted into the same year's new three, four or five year plan, and both are framed in the light of the perspective plan.

In contrast to the rolling plan, there is a fixed plan for four, five, six or seven years. A fixed plan lays down definite aims and objectives which are required to be achieved during the plan period. For this purpose, physical targets are fixed along with the total outlay. Physical targets and financial outlays are seldom changed except under emergencies. Planning in India (Five-Year) and Russia (Seven-Year) is of the fixed type.

CENTRALISED AND DECENTRALISED PLANNING

Under centralized planning, the entire planning process is under a central planning authority. The authority formulates a central plan, fixes objectives, targets, and priorities for every sector of the economy. The principle problems of the economy–what and how much to produce, how and for whom to be produced etc, are decided by this authority. The entire planning process is based on bureaucratic control and regulation. Naturally, such planning is rigid. There is no economic freedom and all economic activities are directed from above. On the other hand, decentralized planning refers to the execution of the plan from the grass roots.

Under it, a plan is formulated by the central planning authority in consultation with the different administrative units of the country. The central plan incorporates plans under the central schemes, and plans for the states under a federal set-up. The state plans incorporate district and village level plans. Under decentralized planning, prices of goods and services are determined by the market mechanism despite government control and regulation in certain fields of economic activity.

ECONOMIC PLANNING

Economic planning refers to any directing or planning of economic activity by the state, in an attempt to achieve specific economic or social outcomes. Planning is an economic mechanism for resource allocation and decision-making in contrast with the market mechanism.

Most economies are mixed economies, incorporating elements of market mechanisms and planning for distributing inputs and outputs. The level of centralization of decision-making ultimately depends on the type of planning mechanism employed; as such planning may be based on either centralized or decentralized decision-making. Economic planning can apply to production, investment, distribution or all three of these functions. Planning may take the form of directive planning or indicative planning. An economy primarily based on central planning is a planned economy; in a planned economy the allocation of resources is determined by a comprehensive plan of production which specifies output requirements.

Socialist Economic Planning

Socialists and Marxists differentiate between the concept of command economy, which existed in the Soviet Union, and economic planning, defining a command economy as top-down administrative planning based on bureaucratic organization. Classical socialists and Marxists define economic planning as directly producing use-values and coordination of production, and consider this to be a fundamental element of a socialist economy. Economic planning implies production for use, planned use in the allocation of surpluses and often calculation-in-kind. For Marxists in particular, planning also entails control of the surplus product by the associated producers in a decentralized, democratic fashion.

Marxists and early technocratic socialists hold the view that in a socialist society based on economic planning, the primary function of the state apparatus will change from one of political rule over people into a scientific administration of things and a direction of processes of production; that is the state would become a coordinating economic entity rather than a mechanism of class or political control, thereby ceasing to be a state in the traditional sense. Libertarian socialists, Syndicalists, Trotskyists, orthodox Marxists and democratic socialists advocate de-centralized participatory planning. In a de-centralized planned economy, economic decision-making is based on self-management and self-governance and a democratic manner from the bottom-up without any directing central authority.

Other socialists, such as Leninists, Marxist-Leninists, Social democrats and some state-oriented socialists advocate directive planning where directives are passed down from higher authorities to agents, who in turn give orders to workers. In some socialist theories, economic planning completely substitutes the market mechanism and supposedly renders monetary relations and the price system obsolete. In other theories, planning is utilized as a complement to markets. Polish economist Oskar Lange and American economist Abba Lerner proposed a form of market socialism where a central planning board would adjust prices of publicly-owned firms to equal marginal cost to enhance the market mechanism by achieving pareto efficient outcomes.

In general, the various types of socialist economic planning exist as theoretical constructs that have not been implemented fully by any economy, partially because they depend on vast changes in social and economic development on a global scale. In the context of mainstream economics, socialist planning usually refers to the Soviet-type command economies, regardless of whether or not they actually constituted a type of state capitalism or a third, non-socialist and non-capitalist system such as Coordinatorism and Bureaucratic collectivism.

Intra-firm and Intra-industry Planning

Large corporations allocate resources internally among different divisions and subsidiaries through planning. Many modern firms also utilize regression analysis to measure market demand in order to adjust price and decide on the optimal quantity of output to be supplied. Planned obsolescence is often cited as a form of economic planning employed by large firms to increase demand for future products by deliberately limiting the operational lifespan of a product. In The New Industrial State, economist John Kenneth Galbraith posited that large firms can manage prices and consumer demand, and because of increasing technological capacity, management has become increasingly specialized and bureaucratized.

The internal structure of a corporation has been reorganized in what he calls a "technostructure", where specialized groups and committees are the primary decision-makers and specialized managers, directors and financial advisers and formal, bureaucratic procedures have replaced the individual entrepreneur's role. He states that both the obsolete notion of "entrepreneurial capitalism" and democratic socialism are impossible for managing the modern industrial system.

Austrian economist Joseph Schumpeter argued that the changing nature of economic activity; the increasing bureaucratization and specialization of production; was one of the main reasons capitalism would eventually evolve into socialism. The role of the businessman was increasingly bureaucratic and specific functions within the firm required increasing specialized knowledge, which can just as easily be supplied by the state apparatus. In the first volume of Capital, Karl Marx identifies a tendency for capital to accumulate under capitalism, which leads to increasing industrial capacity due to increasing returns to scale.

Capitalism eventually socializes labour and production to a point where the traditional notion of private ownership and commodity production are insufficient for managing and further expanding the productive capabilities of society, necessitating a socialist economy of the means of production and cooperative worker control over the surplus value.

Socialists see this as evidence of the increasing obsolescence and inapplicability of notions of perfect competition and as evidence of the increasingly trend towards economic planning in some form or another, the next stage of evolution being planning production on the level of the national economy.

Criticism

The most notable critique of economic planning came from Austrian economists Friedrich Hayek and Ludwig von Mises. Hayek argued that central planners could not possibly accrue the necessary information to formulate an effective plan for production because they are not exposed to the rapid changes in the particular time and place that take place in an economy, and are unfamiliar with these circumstances. Transmitting all the necessary information to planners to accumulate and form a comprehensive plan is therefore inefficient.

Centralized economic planning has also been criticized by proponents of de-centralized economic planning. For example, Leon Trotsky believed that central planners, regardless of their intellectual capacity, operated without the input and participation of the millions of people who participate in the economy and understand/respond to local conditions and changes in the economy would be unable to effectively coordinate all economic activity. Proponents of technocratic planning have responded by saying democratic planning would be inefficient due to the time it takes to deliberate and vote on action in a direct democratic setting. Democratic planning would also be ineffective because various economic decisions require specialized knowledge, which the majority of voters lack.

ENTERPRISE ARCHITECTURE PLANNING

Enterprise Architecture Planning in Enterprise Architecture is the planning process of defining architectures for the use of information in support of the business and the plan for implementing those architectures. One of the earlier professional practitioners in the field of system architecture Steven H. Spewak in 1998 defined Enterprise Architecture Planning as "the process of defining architectures for the use of information in support of the business and the plan for implementing those architectures." Spewak's approach to EAP is similar to that taken by DOE in that the business mission is the primary driver.

That is followed by the data required to satisfy the mission, followed by the applications that are built using that data, and finally by the technology to implement the applications.

In which the layers are implemented in order, from top to bottom. Based on the Business Systems Planning approach developed by John Zachman, EAP takes a data-centric approach to architecture planning to provide data quality, access to data, adaptability to changing requirements, data interoperability and sharing, and cost containment. This view counters the more traditional view that applications should be defined before data needs are determined or provided for.

EAP Topics

Zachman Framework

EAP defines the blueprint for subsequent design and implementation and it places the planning/defining stages into a framework. It does not explain how to define the top two rows of the Zachman Framework in detail but for the sake of the planning exercise, abbreviates the analysis. The Zachman Framework provides the broad context for the description of the architecture layers, while EAP focuses on planning and managing the process of establishing the business alignment of the architectures.

EAP is planning that focuses on the development of matrixes for comparing and analyzing data, applications, and technology. Most important, EAP produces an implementation plan. Within the Federal Enterprise Architecture, EAP will be completed segment enterprise by segment enterprise. The results of these efforts may be of Governmentwide value; therefore, as each segment completes EAP, the results will be published on the ArchitecturePlus web site.

EAP Components

Enterprise Architecture Planning model consists of four levels:

1. *Layer 1 - getting started*: This layer leads to producing an EAP workplan and stresses the necessity of high-level management commitment to support and resource the subsequent six components of the process. It consists of Planning Initiation, which covers in

general, decisions on which methodology to use, who should be involved, what other support is required, and what toolset will be used.

2. *Layer 2 - where we are today:* This layer provides a baseline for defining the eventual architecture and the long-range migration plan. It consists of:
   - Business process modeling, the compilation of a knowledge base about the business functions and the information used in conducting and supporting the various business processes, and
   - Current Systems and Technology, the definition of current application systems and supporting technology platforms.
3. *Layer 3 - the vision of where we want to be*: The arrows delineate the basic definition process flow: data architecture, applications architecture, and technology architecture. It consists of:
   - *Data Architecture*: Definition of the major kinds of data needed to support the business.
   - *Applications Architecture*: Definition of the major kinds of applications needed to manage that data and support the business functions.
   - *Technology Architecture*: Definition of the technology platforms needed to support the applications that manage the data and support the business functions.
4. *Layer 4 - how we plan to get there*: This consists of the Implementation/ Migration Plans - Definition of the sequence for implementing applications, a schedule for implementation, a cost/benefit analysis, and a clear path for migration.

EAP Methodology

The Enterprise Architecture Planning methodology is beneficial to understanding the further definition of the Federal Enterprise Architecture Framework at level IV. EAP is a how to approach for creating the top two rows of the Zachman Framework, Planner and Owner. The design of systems begins in the third row, outside the scope of EAP. EAP focuses on defining what data, applications, and technology architectures are appropriate for and support the overall enterprise. The seven components of EAP for defining these architectures and the related migration plan. The seven components are in the shape of a wedding cake, with each layer representing a different focus of each major task.

EVENT PLANNING

Event planning is the process of planning a festival, ceremony, competition, party, or convention. Event planning includes budgeting, establishing dates and alternate dates, selecting and reserving the event site, acquiring permits, and coordinating transportation and parking. Event planning also includes some or all of the following, depending on the event: developing a theme or motif for the event, arranging for speakers and alternate speakers, coordinating location support arranging decor, tables, chairs, tents, event support and security, catering, police, fire, portable toilets, parking, signage, emergency plans, health care professionals, and cleanup.

Steps to Planning an Event

The first step to planning an event is determining its purpose, whether it is for a wedding, company, birthday, festival, graduation or any other event requiring extensive planning.

From this the event planner needs to choose entertainment, location, guest list, speakers, and content. The location for events is endless, but with event planning they would likely be held at hotels, convention centers, reception halls, or outdoors depending on the event. Once the location is set the coordinator/ planner needs to prepare the event with staff, set up the entertainment, and keep contact with the client.

After all this is set the event planner has all the smaller details to address like set up of the event such as food, drinks, music, guest list, budget, advertising and marketing, decorations, all this preparation is what is needed for an event to run smoothly. An event planner needs to be able to manage their time wisely for the event, and the length of preparation needed for each event so it is a success.

Event Planning as a Career

Event planning is a relatively new career field. There is now training that helps one trying to break into the career field. There must be training for an event planner to handle all the pressure and work efficiently. This career deals with a lot of communication and organization aspects.

There are many different names for an event planner such as a conference coordinator, a convention planner, a special event coordinator, and a meeting manager.

To read more about people involved in event planning and production. Event planners' work is considered either stressful or energizing. This line of work is also considered fast paced and demanding. Planners face deadlines and communicating with multiple people at one time. Planners spend most of their time in offices, but when meeting with clients the work is usually on-site at the location where the event is taking place.

Some physical activity is required such as carrying boxes of materials and decorations or supplies needed for the event. Also, long working hours can be a part of the job. The day the event is taking place could start as early as 5:00 a.m. and then work until midnight. Working on weekends is sometimes required, which is when many events take place.

FINANCIAL PLAN

In general usage, a financial plan can be a budget, a plan for spending and saving future income. This plan allocates future income to various types of expenses, such as rent or utilities, and also reserves some income for short-term and long-term savings.

A financial plan can also be an investment plan, which allocates savings to various assets or projects expected to produce future income, such as a new business or product line, shares in an existing business, or real estate. In business, a financial plan can refer to the three primary financial statements created within a business plan. Financial forecast or financial plan can also refer to an annual projection of income and expenses for a company, division or department. A financial plan can also be an estimation of cash needs and a decision on how to raise the cash, such as through borrowing or issuing additional shares in a company. While a financial plan refers to estimating future income, expenses and assets, a financing plan or finance plan usually refers to the means by which cash will be acquired to cover future expenses, for instance through earning, borrowing or using saved cash.

MARKETING PLAN

A marketing plan may be part of an overall business plan. Solid marketing strategy is the foundation of a well-written marketing plan. While a marketing plan contains a list of actions, a marketing plan without a sound strategic foundation is of little use.

The Marketing Planning Process

Marketing process can be realised by the marketing mix in step 4. The last step in the process is the marketing controlling. In most organizations, "strategic planning" is an annual process, typically covering just the year ahead. Occasionally, a few organizations may look at a practical plan which stretches three or more years ahead. To be most effective, the plan has to be formalized, usually in written form, as a formal "marketing plan." The essence of the process is that it moves from the general to the specific, from the vision to the mission to the goals to the corporate objectives of the organization, then down to the individual action plans for each part of the marketing programme. It is also an interactive process, so that the draft output of each stage is checked to see what impact it has on the earlier stages, and is amended.

Marketing Planning Aims and Objectives

Behind the corporate objectives, which in themselves offer the main context for the marketing plan, will lie the "corporate mission," which in turn provides the context for these corporate objectives. In a sales-oriented organization, the marketing planning function designs incentive pay plans to not only motivate and reward frontline staff fairly but also to align marketing activities with corporate mission.

This "corporate mission" can be thought of as a definition of what the organization is, of what it does: "Our business is ...". This definition should not be too narrow, or it will constrict the development of the organization; a too rigorous concentration on the view that "We are in the business of making meat-scales," as IBM was during the early 1900s, might have limited its subsequent development into other areas. On the other hand, it should not be too wide or it will become meaningless; "We want to make a profit" is not too helpful in developing specific plans. Abell suggested that the definition should cover three dimensions: "customer groups" to be served, "customer needs" to be served, and "technologies" to be used. Thus, the definition of IBM's "corporate mission" in the 1940s might well have been: "We are in the business of handling accounting information for the larger US organizations by means of punched cards." Perhaps the most important factor in successful marketing is the "corporate vision."

Surprisingly, it is largely neglected by marketing textbooks, although not by the popular exponents of corporate strategy - indeed, it was perhaps the main theme of the book by Peters and Waterman, in the form of their "Superordinate Goals." "In Search of Excellence" said: "Nothing drives progress like the imagination. The idea precedes the deed." If the organization in general, and its chief executive in particular, has a strong vision of where its future lies, then there is a good chance that the organization will achieve a strong position in its markets. This will be not least because its strategies will be consistent and will be supported by its staff at all levels. In this context, all of IBM's marketing activities were underpinned by its philosophy of "customer service," a vision originally promoted by the charismatic Watson dynasty. The emphasis at this stage is on obtaining a complete and accurate picture.

A "traditional" - albeit product-based - format for a "brand reference book" was suggested by Godley more than three decades ago:

- *Financial data*: Facts for this part will come from management accounting, costing and finance parts.
- *Product data*: From production, research and development.
- *Sales and distribution data*: Sales, packaging, distribution parts.
- *Advertising, sales promotion, merchandising data*: Information from these departments.
- *Market data and miscellany*: From market research, who would in most cases act as a source for this information. His sources of data, however, assume the resources of a very large organization. In most organizations they would be obtained from a much smaller set of people.

It is apparent that a marketing audit can be a complex process, but the aim is simple: "it is only to identify those existing factors which will have a significant impact on the future plans of the company." It is clear that the basic material to be input to the marketing audit should be comprehensive. The best approach is to accumulate this material continuously, as and when it becomes available; since this avoids the otherwise heavy workload involved in collecting it as part of the regular, typically annual, planning process itself - when time is usually at a premium. Even so, the first task of this annual process should be to check that the material held in the current facts book or facts files actually is comprehensive and accurate, and can form a sound basis for the marketing audit itself. The structure of the facts book will be designed to match the specific needs of the organization, but one simple format - suggested by Malcolm McDonald - may be applicable in many cases.

This splits the material into three groups:

1. Review of the marketing environment. A study of the organization's markets, customers, competitors and the overall economic, political, cultural and technical environment; covering developing trends, as well as the current situation.
2. Review of the detailed marketing activity. A study of the company's marketing mix; in terms of the 7 Ps
3. Review of the marketing system. A study of the marketing organization, marketing research systems and the current marketing objectives and strategies. The last of these is too frequently ignored. The marketing system itself needs to be regularly questioned, because the validity of the whole marketing plan is reliant upon the accuracy of the input from this system, and 'garbage in, garbage out' applies with a vengeance.
   - *Portfolio planning*: In addition, the coordinated planning of the individual products and services can contribute towards the balanced portfolio.
   - 80:20 rule. To achieve the maximum impact, the marketing plan must be clear, concise and simple. It needs to concentrate on the 20 per cent of products or services, and on the 20 per cent of customers, that will account for 80 per cent of the volume and 80 per cent of the profit.
   - 7 Ps: Product, Place, Price and Promotion, Physical Environment, People, Process. The 7 Ps can sometimes divert attention from the customer, but the framework they offer can be very useful in building the action plans.

It is only at this stage that the active part of the marketing planning process begins. This next stage in marketing planning is indeed the key to the whole marketing process. The "marketing objectives" state just where the company intends to be at some specific time in the future. James Quinn succinctly defined objectives in general as: Goals state what is to be achieved and when results are to be accomplished, but they do not state "how" the results are to be achieved. They typically relate to what products will be where in what markets. They are essentially about the match between those "products" and "markets."

Objectives for pricing, distribution, advertising and so on are at a lower level, and should not be confused with marketing objectives. They are part of the marketing strategy needed to achieve marketing objectives. To be most effective, objectives should be capable of measurement and therefore "quantifiable." This measurement may be in terms of sales volume, money value, market share, percentage penetration of distribution outlets and so on. An example of such a measurable marketing objective might be "to enter the market with product Y and capture 10 per cent of the market by value within one year."

As it is quantified it can, within limits, be unequivocally monitored, and corrective action taken as necessary. The marketing objectives must usually be based on the organization's financial objectives; converting these financial measurements into the related marketing measurements.He went on to explain his view of the role of "policies," with which strategy is most often confused: "Policies are rules or guidelines that express the 'limits' within which action should occur."Simplifying somewhat, marketing strategies can be seen as the means, or "game plan," by which marketing objectives will be achieved and, in the framework that we have chosen to use, are generally concerned with the 8 P's.

Examples are:

1. *Price*: The amount of money needed to buy products
2. *Product*: The actual product
3. *Promotion*: Getting the product known
4. *Placement*: Where the product is located
5. *People*: Represent the business
6. *Physical environment*: The ambiance, mood, or tone of the environment
7. *Process*: How do people obtain your product
8. *Packaging*: How the product will be protected

In principle, these strategies describe how the objectives will be achieved. The 7 Ps are a useful framework for deciding how the company's resources will be manipulated to achieve the objectives. However, they are not the only framework, and may divert attention from the real issues. The focus of the strategies must be the objectives to be achieved - not the process of planning itself. Only if it fits the needs of these objectives should you choose, as we have done, to use the framework of the 7 Ps. The strategy statement can take the form of a purely verbal description of the strategic options which have been chosen.

Alternatively, and perhaps more positively, it might include a structured list of the major options chosen. One aspect of strategy which is often overlooked is that of "timing." Exactly when it is the best time for each element of the strategy to be implemented is often critical. Taking the right action at the wrong time can sometimes be almost as bad as taking the wrong action at the right time.

Timing is, therefore, an essential part of any plan; and should normally appear as a schedule of planned activities.Having completed this crucial stage of the planning process, you will need to re-check the feasibility of your objectives and strategies in terms of the market share, sales, costs, profits and so on which these demand in practice. As in the rest of the marketing discipline, you will need to employ judgement, experience, market research or anything else which helps you to look at your conclusions from all possible angles.

Detailed Plans and Programmes

At this stage,you will need to develop your overall marketing strategies into detailed plans and programme. Although these detailed plans may cover each of the 7 Ps, the focus will vary, depending upon your organization's specific strategies. A product-oriented company will focus its plans for the 7 Ps around each of its products. A market or geographically oriented company will concentrate on each market or geographical area.

Each will base its plans upon the detailed needs of its customers, and on the strategies chosen to satisfy these needs. Brochures and Websites are used effectively. Again, the most important element is, indeed, that of the detailed plans, which spell out exactly what programmes and individual activities will take place over the period of the plan. Without these specified - and preferably quantified - activities the plan cannot be monitored, even in terms of success in meeting its objectives.It is these programmes and activities which will then constitute the "marketing" of the organization over the period. As a result, these detailed marketing programmes are the most important, practical outcome of the whole planning process.

These plans must therefore be:

- *Clear*: They should be an unambiguous statement of 'exactly' what is to be done.
- *Quantified*: The predicted outcome of each activity should be, as far as possible, quantified, so that its performance can be monitored.
- *Focused*: The temptation to proliferate activities beyond the numbers which can be realistically controlled should be avoided. The 80:20 Rule applies in this context too.
- *Realistic*: They should be achievable.
- *Agreed*: Those who are to implement them should be committed to them, and agree that they are achievable. The resulting plans should become a working document which will guide the campaigns taking place throughout the organization over the period of the plan. If the marketing plan is to work, every exception to it must be questioned; and the lessons learnt, to be incorporated in the next year's.

Measurement of Progress

The final stage of any marketing planning process is to establish targets so that progress can be monitored. It is important to put both quantities and timescales into the marketing objectives and into the corresponding strategies. Changes in the environment mean that the forecasts often have to be changed. Along with these, the related plans may well also need to be changed. Continuous monitoring of performance, against predetermined targets, represents a most important aspect of this. However, perhaps even more important is the enforced discipline of a regular formal review. Again, as with forecasts, in many cases the best planning cycle will revolve around a quarterly review. Best of all, at least in terms of the quantifiable aspects of the plans, if not the wealth of backing detail, is probably a quarterly rolling review - planning one full year ahead each new quarter. Of course, this does absorb more planning resource; but it also ensures that the plans embody the latest information, and - with attention focused on them so regularly - forces both the plans and their implementation to be realistic. Plans only have validity if they are actually used to control the progress of a company: their success lies in their implementation, not in the writing'.

Performance Analysis

The most important elements of marketing performance, which are normally tracked, are:

Sales Analysis

Most organizations track their sales results; or, in non-profit organizations for example, the number of clients. The more sophisticated track them in terms of 'sales variance' - the deviation from the target figures - which allows a more immediate picture of deviations to become evident. `Micro-analysis', which is simply the normal management process of investigating detailed problems, then investigates the individual elements which are failing to meet targets.

Market Share Analysis

Few organizations track market share though it is often an important metric. Though absolute sales might grow in an expanding market, a firm's share of the market can decrease which bodes ill for future sales when the market starts to drop.

Where such market share is tracked, there may be a number of aspects which will be followed:

- Overall market share
- Segment share - that in the specific, targeted segment
- Relative share -in relation to the market leaders
- Annual fluctuation rate of market share
- also the specific market sharing of customers.

Expense Analysis

The key ratio to watch in this area is usually the 'marketing expense to sales ratio'; although this may be broken down into other elements.

Financial Analysis

The "bottom line" of marketing activities should at least in theory, be the net profit.

There are a number of separate performance figures and key ratios which need to be tracked:

- Gross contribution< >net profit
- Gross profit< >return on investment
- Net contribution< >profit on sales

There can be considerable benefit in comparing these figures with those achieved by other organizations; using, for instance, the figures which can be obtained from 'The Center for Interfirm Comparison'. The most sophisticated use of this approach, however, is typically by those making use of PIMS, initiated by the General Electric Company and then developed by Harvard Business School, but now run by the Strategic Planning Institute.

The performance analyses concentrate on the quantitative measures which are directly related to short-term performance. But there are a number of indirect measures, essentially tracking customer attitudes, which can also indicate the organization's performance in terms of its longer-term marketing strengths and may accordingly be even more important indicators.

Some useful measures are:

- *Market research*: Including customer panels lost business–the orders which were lost because, for example, the stock was not available or the product did not meet the customer's exact requirements
- *Customer complaints*: How many customers complain about the products or services, or the organization itself, and about what

Use of Marketing Plans

A formal, written marketing plan is essential; in that it provides an unambiguous reference point for activities throughout the planning period. However, perhaps the most important benefit of these plans is the planning process itself. This typically offers a unique opportunity, a forum, for information-rich and productively focused discussions between the various managers involved.

The plan, together with the associated discussions, then provides an agreed context for their subsequent management activities, even for those not described in the plan itself. Additionally, marketing plans are included in business plans, offering data showing investors how the company will grow and most importantly, how they will get a return on investment.

Budgets as Managerial Tools

The classic quantification of a marketing plan appears in the form of budgets. Because these are so rigorously quantified, they are particularly important. They should, thus, represent an unequivocal projection of actions and expected results. What is more, they should be capable of being monitored accurately; and, indeed, performance against budget is the main management review process.

The purpose of a marketing budget is, thus, to pull together all the revenues and costs involved in marketing into one comprehensive document. It is a managerial tool that balances what is needed to be spent against what can be afforded, and helps make choices about priorities. It is then used in monitoring performance in practice.

The marketing budget is usually the most powerful tool by which you think through the relationship between desired results and available means. Its starting point should be the marketing strategies and plans, which have already been formulated in the marketing plan itself; although, in practice, the two will run in parallel and will interact. At the very least, the rigorous, highly quantified, budgets may cause a rethink of some of the more optimistic elements of the plans.

NETWORK RESOURCE PLANNING

Network Resource Planning is an enhanced process of network planning that incorporates the disciplines of business planning, marketing, and engineering to develop integrated, dynamic master plans for all domains of communications networks.

SUCCESSION PLANNING

Succession planning is a process for identifying and developing internal people with the potential to fill key leadership positions in the company. Succession planning increases the availability of experienced and capable employees that are prepared to assume these roles as they become available. Taken narrowly, "replacement planning" for key roles is the heart of succession planning.

Effective succession or talent-pool management concerns itself with building a series of feeder groups up and down the entire leadership pipeline or progression. In contrast, replacement planning is focused narrowly on identifying specific back-up candidates for given senior management positions. For the most part position-driven replacement planning is a forecast, which research indicates does not have substantial impact on outcomes. Fundamental to the succession-management process is an underlying philosophy that argues that top talent in the corporation must be managed for the greater good of the enterprise.

Merck and other companies argue that a "talent mindset" must be part of the leadership culture for these practices to be effective. Research indicates many succession-planning initiatives fall short of their intent. "Bench strength," as it is commonly called, remains a stubborn problem in many if not most companies. Studies indicate that companies that report the greatest gains from succession planning feature high ownership by the CEO and high degrees of engagement among the larger leadership team Companies that are well known for their succession planning and executive talent development practices include: GE, Honeywell, IBM, Marriott, Microsoft, Pepsi and Proctor and Gamble. Research indicates that clear objectives are critical to establishing effective succession planning.

These objectives tend to be core to many or most companies that have well-established practices:

- Identify those with the potential to assume greater responsibility in the organization
- Provide critical development experiences to those that can move into key roles
- Engage the leadership in supporting the development of high-potential leaders
- Build a data base that can be used to make better staffing decisions for key jobs

In other companies these additional objectives may be embedded in the succession process:

- Improve employee commitment and retention
- Meet the career development expectations of existing employees
- Counter the increasing difficulty and costs of recruiting employees externally

Field of Succession Management

There is a substantial body of literature on the subject of succession planning. The first book that addressed the topic fully was "Executive Continuity" by Walter Mahler. Mahler was responsible in the 1970s for helping to shape the General Electric succession process which became the gold standard of corporate practice.

Mahler, who was heavily influenced by Peter Drucker, wrote three other books on the subject of succession, all of which are out of print. His colleagues, Steve Drotter and Greg Kesler, as well as others, expanded on Mahler's work in their writings. "The Leadership Pipeline: How to Build the Leadership Powered Company," by Charan, Drotter and Noel is noteworthy. A new edited collection of materials, edited by Marshall Goldsmith, describes many contemporary examples in large companies.

Most large corporations assign a process owner for talent and succession management. Resourcing of the work varies widely from numbers of highly dedicated internal consultants to limited professional support embedded in the roles of human resources generalists. Often these staff resources are separate from external staffing or recruiting functions. Some companies today seek to integrate internal and external staffing. Others are more inclined to integrate succession management with the performance management process in order simplify the work for line managers.

Large consultancies, such as McKinsey, have recently focused on the broader talent issue, but most consulting support to executive succession and development practices probably comes from numerous boutique firms and retired executives in the field. The leading professional affiliation for succession-planning professionals is, arguably, The Human Resources Planning Society, an international association with academic and practitioner membership from around the world.

Process and Practices

Companies devise elaborate models to characterize their succession and development practices.

Most reflect a cyclical series of activities that include these fundamentals:

- Identify key roles for succession or replacement planning
- Define the competencies and motivational profile required to undertake those roles
- Assess people against these criteria - with a future orientation
- Identify pools of talent that could potentially fill and perform highly in key roles
- Develop employees to be ready for advancement into key roles - primarily through the right set of experiences.

In many companies, over the past several years, the emphasis has shifted from planning job assignments to development, with much greater focus on managing key experiences that are critical to growing global business leaders. North American companies tend to be more active in this regard, followed by European and Latin American countries. PepsiCo, IBM and Nike are current examples of the so-called "game planning" approach to succession and talent management. In these and other companies annual reviews are supplemented with an ongoing series of discussions among senior leaders about who is ready to assume larger roles. Vacancies are anticipated and slates of names are prepared based on highest potential and readiness for job moves.

Organization realignments are viewed as critical windows of opportunity to create development moves that will serve the greater good of the enterprise. Assessment is a key practice in effective succession planning. There is no widely accepted formula for evaluating the future potential of leaders, but there are many tools and approaches that continue to be used today, ranging from personality and cognitive testing to team-based interviewing and simulations and other assessment center methods.

Eliot Jaques and others have argued for the importance of focusing assessments narrowly on critical differentiators of future performance. Jaques developed a persuasive case for measuring candidates' ability to manage complexity, a robust, but operational definition of business intelligence. Companies struggle to find practices that are effective and practical. It is clear leaders who rely on instinct and gut to make promotion decisions are often not effective.

Research indicates that the most valid practices for assessment are those that involve multiple methods and especially multiple raters "Calibration meetings," composed of senior leaders can be quite effective judging a slate of potential senior leaders with the right tools and facilitation. Professionals in the field, including academics, consultants and corporate practitioners, have many strongly held views on the topic. Best practice is a slippery concept in this field.

There are many thought pieces on the subject that readers may find valuable such as "Debunking 10 Top Talent Management Myths", Talent Management Magazine, Doris Sims, December 2009. Research-based writing is more difficult to find. The Corporate Leadership Council, The Best Practice Institute and the Center for Creative Leadership, as well as the Human Resources Planning Society are sources of some effective research-based materials.

DECISION MAKING: CONCEPT AND PROCESS

All of us have to make decisions every day. Some decisions are relatively straightforward and simple: Is this report ready to send to my boss now? Others are quite complex: Which of these candidates should I select for the job? Simple decisions usually need a simple decision-making process.

But difficult decisions typically involve issues like these:

- *Uncertainty*: Many facts may not be known.
- *Complexity*: You have to consider many interrelated factors.
- *High-risk consequences:* The impact of the decision may be significant.
- *Alternatives:* Each has its own set of uncertainties and consequences.
- *Interpersonal issues*: It can be difficult to predict how other people will react.

With these difficulties in mind, the best way to make a complex decision is to use an effective process. Clear processes usually lead to consistent, high-quality results, and they can improve the quality of almost everything we do.

A SYSTEMATIC APPROACH TO DECISION MAKING

A logical and systematic decision-making process helps you address the critical elements that result in a good decision. By taking an organized approach, you're less likely to miss important factors, and you can build on the approach to make your decisions better and better.

There are six steps to making an effective decision:

1. Create a constructive environment.
2. Generate good alternatives.
3. Explore these alternatives.
4. Choose the best alternative.
5. Check your decision.
6. Communicate your decision, and take action.

Create a Constructive Environment

To create a constructive environment for successful decision making, make sure you do the following:

- *Establish the objective*: Define what you want to achieve.
- *Agree on the process*: Know how the final decision will be made, including whether it will be an individual or a team-based decision. The Vroom-Yetton-Jago Model is a great tool for determining the most appropriate way of making the decision.
- *Involve the right people*: Stakeholder Analysis is important in making an effective decision, and you'll want to ensure that you've consulted stakeholders appropriately even if you're making an individual decision. Where a group process is appropriate, the decision-making group-typically a team of five to seven people-should have a good representation of stakeholders.
- *Allow opinions to be heard:* Encourage participants to contribute to the discussions, debates, and analysis without any fear of rejection from the group. This is one of the best ways to avoid groupthink. The Stepladder Technique is a useful method for gradually introducing more and more people to the group discussion, and making sure everyone is heard. Also, recognize that the objective is to make the best decision under the circumstances: it's not a game in which people are competing to have their own preferred alternatives adopted.
- *Make sure you're asking the right question*: Ask yourself whether this is really the true issue. The 5 Whys technique is a classic tool that helps you identify the real underlying problem that you face.
- *Use creativity tools from the start*: The basis of creativity is thinking from a different perspective. Do this when you first set out the problem, and then continue it while generating alternatives.

Generate Good Alternatives

This step is still critical to making an effective decision. The more good options you consider, the more comprehensive your final decision will be. When you generate alternatives, you force yourself to dig deeper, and look at the problem from different angles. If you use the mindset 'there must be other solutions out there,' you're more likely to make the best decision possible. If you don't have reasonable alternatives, then there's really not much of a decision to make! *Here's a summary of some of the key tools and techniques to help you and your team develop good alternatives*:

- Generating Ideas
- Brainstorming is probably the most popular method of generating ideas.
- Another approach, Reverse Brainstorming, works similarly. However, it starts by asking people to brainstorm how to achieve the opposite outcome from the one wanted, and then reversing these actions.
- The Charette Procedure is a systematic process for gathering and developing ideas from very many stakeholders.
- Use the Crawford Slip Writing Technique to generate ideas from a large number of people. This is an extremely effective way to make sure that everyone's ideas are heard and given equal weight, irrespective of the person's position or power within the organization.
- Considering Different Perspectives
- The Reframing Matrix uses 4 Ps as the basis for gathering different perspectives. You can also ask outsiders to join the discussion, or ask existing participants to adopt different functional perspectives.
- If you have very few options, or an unsatisfactory alternative, use a Concept Fan to take a step back from the problem, and approach it from a wider perspective. This often helps when the people involved in the decision are too close to the problem.
- Appreciative Inquiry forces you to look at the problem based on what's 'going right,' rather than what's 'going wrong.'
- Organizing Ideas
- This is especially helpful when you have a large number of ideas. Sometimes separate ideas can be combined into one comprehensive alternative.
- Use Affinity Diagrams to organize ideas into common themes and groupings.

Explore the Alternatives

When you're satisfied that you have a good selection of realistic alternatives, then you'll need to evaluate the feasibility, risks, and implications of each choice. Here, we discuss some of the most popular and effective analytical tools.

- Risk
- In decision making, there's usually some degree of uncertainty, which inevitably leads to risk. By evaluating the risk involved with various options, you can determine whether the risk is manageable.
- Risk Analysis helps you look at risks objectively. It uses a structured approach for assessing threats, and for evaluating the probability of events occurring - and what they might cost to manage.

- Implications
- Another way to look at your options is by considering the potential consequences of each.
- Six Thinking Hats helps you evaluate the consequences of a decision by looking at the alternatives from six different perspectives.
- Impact Analysis is a useful technique for brainstorming the 'unexpected' consequences that may arise from a decision.
- Validation
- Determine if resources are adequate, if the solution matches your objectives, and if the decision is likely to work in the long term.
- Starbursting helps you think about the questions you should ask to evaluate an alternative properly.
- To assess pros and cons of each option, use Force Field Analysis, or use the Plus-Minus-Interesting approach.
- Cost-Benefit Analysis looks at the financial feasibility of an alternative.
- Our Bite-Sized Training session on Project Evaluation and Financial Forecasting helps you evaluate each alternative using the most popular financial evaluation techniques.

Choose the Best Alternative

After you have evaluated the alternatives, the next step is to choose between them. The choice may be obvious.

However, if it isn't, these tools will help:

- Grid Analysis, also known as a decision matrix, is a key tool for this type of evaluation. It's invaluable because it helps you bring disparate factors into your decision-making process in a reliable and rigorous way.
- Use Paired Comparison Analysis to determine the relative importance of various factors. This helps you compare unlike factors, and decide which ones should carry the most weight in your decision.
- Decision Trees are also useful in choosing between options. These help you lay out the different options open to you, and bring the likelihood of project success or failure into the decision making process.

Check Your Decision

With all of the effort and hard work that goes into evaluating alternatives, and deciding the best way forward, it's easy to forget to 'sense check' your decisions. This is where you look at the decision you're about to make dispassionately, to make sure that your process has been thorough, and to ensure that common errors haven't crept into the decision-making process.

After all, we can all now see the catastrophic consequences that over-confidence, groupthink, and other decision-making errors have wrought on the world economy. The first part of this is an intuitive step, which involves quietly and methodically testing the assumptions and the decisions you've made against your own experience, and thoroughly reviewing and exploring any doubts you might have.

A second part involves using a technique like Blindspot Analysis to review whether common decision-making problems like over-confidence, escalating commitment, or groupthink may have undermined the decision-making process. A third part involves using a technique like the Ladder of Inference to check through the logical structure of the decision with a view to ensuring that a well-founded and consistent decision emerges at the end of the decision-making process.

Communicate Your Decision, and Move to Action!

Once you've made your decision, it's important to explain it to those affected by it, and involved in implementing it. Talk about why you chose the alternative you did. The more information you provide about risks and projected benefits, the more likely people are to support the decision. And with respect to implementation of your decision, our objects on Project Management and Change Management will help you get this implementation off to a good start!

MANAGEMENT DECISION MAKING

A major concern in management has been to understand and improve decision making. Various approaches have been proposed by psychologists, most based on a "divide-and-conquer" strategy. This strategy–also labeled "problem decomposition"–involves breaking a large decision problem into smaller parts. The idea is not new: In a "Letter to Joseph Priestly," Benjamin Franklin was one of the first to describe a decomposition strategy. The theoretical justification for this approach was outlined by Simon in his account of "bounded rationality."

This concept says that cognitive processing limitations leave humans with little option but to construct simplified mental models of the world. As Simon put it, a person "behaves rationally with respect to this model... such behaviour is not even approximately optimal with respect to the real world." There have been two approaches to management decision making. The first is concerned with development and application of normative decision rules based on formal logic derived from economics or statistics. The second involves descriptive accounts of how people actually go about making judegments, decisions, and choices.

Normative Analyses

As initially outlined by von Neumann and Morgenstern in Theory of games and economic behaviour, a variety of techniques have been derived for making optimal decisions. A distinction is often drawn between riskless choices and risky choices. Two examples of each approach are outlined here.

Certain Outcomes

- *Multi-Attribute Utility*: This approach, abbreviated MAU, applies to decisions made with moreor-less certain outcomes. As described by Gardiner and Edwards, MAU involves obtaining a utility value for each decision alternative and then selecting the alternative with the highest value. The utility for an alternative is derived from a weighted sum of separate part utilities for various attributes. The MAU approach has been successfully applied to management decisions such as new plant sitings, personnel selection, and zoning decisions.
- *Linear Models*: Initially based on multiple-regression analyses, linear models have been used both to prescribe and to describe judegments under certainty. A major concern has been the weights assigned to the cue values. Research has shown that equal weights, or even random weights, do as well as optimal weights in many settings. The robustness of linear models has allowed their use in many applied tasks, such as graduate school admissions, clinical diagnosis, and medical decision making.

Uncertain Outcomes

- *Decision-Tree Analysis*: "A decision tree is a graphical model that displays the sequence of decisions and the events that comprise a sequential decision situation". The approach involves laying out choice alternatives, uncertain events, and outcome utilities as a series of branches. For each alternative, an expected value is computed as the average outcome value over all possible events. The optimal choice is then the alternative with the highest EV. Decision trees have been used to guide risky decision making such as marketing strategy, plant expansion, and public policy planning.
- *Bayesian Networks*: This approach combines elements of Bayesian probability theory, artificial intelligence, and graphical analysis into a decision analytic tool. Starting with a "fully connected" network, all possible cause -and-effect linkages between nodes for a problem space are described. Through a process of "pruning" using computer algorithms, the structure of the network is simplified to essential links between nodes. This results in an enormous reduction of problem complexity. The approach is being used to diagnose computer programming errors and to anticipate trouble spots around the world.

Descriptive Analyses

Most, but not all, descriptive analyses of decision making were initially concerned with accounting for the discrepancies between normative rules and actual behaviour. For instance, Edwards modified EV by substituting subjective probabilities for objective probabilities and psychological utilities for payoff amounts to produce Subjectively Expected Utility. This model has become the starting point for descriptions of risky decision behaviour. However, many other approaches have been offered by psychologists.

Social Judegment Theory

Based on the "Lens Model" proposed by Brunswik, Hammond developed a comprehensive perspective on judegment and decision making. By adapting procedures from multiple regression, this approach combines elements of both normative and descriptive analyses into a single framework. Central to SJT is the distinction between analytic and intuitive modes of cognition. The approach has been used to describe decisions by highway engineers and medical doctors.

Information Integration Theory (IIT)

Analyses of the psychological combination rules used to combine information from multiple sources reveals that people often average stimulus inputs when making judegments. Anderson has shown repeatedly that an averaging rule is more descriptive than the adding or summing rule assumed in normative models.

Through Functional Measurement, IIT leads to the simultaneous evaluation of processing strategy and psychological values. The IIT approach has been applied to marketing decisions, family choices, and expert judegments.

Image Theory

As described by Beach, "image th eory views the decision maker as possessing three distinct but related images, each of which comprise a particular part of his or her decision-related knowledge". The value image consists of the decision maker's values, beliefs, and ethics that collectively are labeled principles. The trajectory image consists of the decision maker's future agenda. And the strategic image consists of various plans that have been adopted to achieve the goals. Using these concepts, image theory has been applied to auditing, childbearing, and political decisions.

Heuristics and Biases

Tversky and Kahneman have argued that decisions are often made using psychological shortcuts or "heuristics." For instance, the "representativeness" heuristic refers to a tendency to make probability judegments based on the similarity of an event to an underlying source; the greater the similarity, the higher the probability estimate. Although easy to do psychologically, such heuristics often lead to "biases" in that releva nt information, such as base rates, may be ignored. This approach has been used to account for suboptimal decisions in accounting, management, and marketing.

Fast and Frugal Heuristics

Simon developed "bounded rationality" to deal with two interloc king components: the limitations of the human mind, and the structure of the environment in which humans operate. For instance, "satisficing" is a cogn itively simple, but often surprisingly efficient decision strategy. These ideas have been extended by Gigerenzer and Todd to apply to various simple "fast and frugal" heuristics that take advantage of environmental constraints. These heuristics have been applied to help decision making in medicine and forecasting.

Naturalistic Decision Making

This perspective was developed by Klein to account for on-line decision making by experts in time-sensitive environments. In situations such as fire fighting, there is not enough time to apply normative choice rules. Instead, experienced decision makers frequently follow a "re cognition- primed decision making" strategy–they identify a single course of action through pattern matching. The NDM approach has been applied in many real-world decisions, ranging from military commands and intelligence analysis to medical diagnosis and accounting.

Expert Decision Making

Behind much of the advances in decision research has been the need for psychologists to help professionals make better decisions. For instance, considerable effort has been extended to understand how clinical psychologists make decisions. Although such analyses often reveal that experts are biased in their decisions, there are many domains in which surprisingly good decisions have been observed. For example, weather forecasts were reported by Stewart, to make reliable and valid short-term predictions of precipitation and temperature. Similarly, auditors were found by Krogstad, to have an effective grasp of what information to use in assessing the accuracy of accounting statements.

PROBLEM ANALYSIS VS DECISION MAKING

It is important to differentiate between problem analysis and decision making. The concepts are completely separate from one another. Problem analysis must be done first, then the information gathered in that process may be used towards decision making.

Problem Analysis:

- Analyse performance, what should the results be against what they actually are
- Problems are merely deviations from performance standards
- Problem must be precisely identified and described
- Problems are caused by some change from a distinctive feature

- Something can always be used to distinguish between what has and hasn't been effected by a cause
- Causes to problems can be deducted from relevant changes found in analyzing the problem
- Most likely cause to a problem is the one that exactly explains all the facts

Decision Making:

- Objectives must first be established
- Objectives must be classified and placed in order of importance
- Alternative actions must be developed
- The alternative must be evaluated against all the objectives
- The alternative that is able to achieve all the objectives is the tentative decision
- The tentative decision is evaluated for more possible consequences
- The decisive actions are taken, and additional actions are taken to prevent any adverse consequences from becoming problems and starting both systems all over again
- There are steps that are generally followed that result in a decision model that can be used to determine an optimal production plan.

EVERYDAY TECHNIQUES

Some of the decision making techniques people use in everyday life include:

- *Pros and Cons*: Listing the advantages and disadvantages of each option, popularized by Plato and Benjamin Franklin
- *Simple Prioritization*: Choosing the alternative with the highest probability-weighted utility for each alternative
- *Satisficing*: Using the first acceptable option found
- Acquiesce to a person in authority or an "expert", just following orders
- *Flipism*: Flipping a coin, cutting a deck of playing cards, and other random or coincidence methods
- Prayer, tarot cards, astrology, augurs, revelation, or other forms of divination

DECISION-MAKING STAGES

Developed by B. Aubrey Fisher, there are four stages that should be involved in all group decision making. These stages, or sometimes called phases, are important for the decision-making process to begin

- *Orientation stage*: This phase is where members meet for the first time and start to get to know each other.
- *Conflict stage*: Once group members become familiar with each other, disputes, little fights and arguments occur. Group members eventually work it out.

- *Emergence stage*: The group begins to clear up vague opinions by talking about them.
- *Reinforcement stage*: Members finally make a decision, while justifying themselves that it was the right decision.

DECISION-MAKING STEPS

When in an organization and faced with a difficult decision, there are several steps one can take to ensure the best possible solutions will be decided. These steps are put into seven effective ways to go about this decision making process.

- *The first step*: Outline your goal and outcome. This will enable decision makers to see exactly what they are trying to accomplish and keep them on a specific path.
- *The second step*: Gather data. This will help decision makers have actual evidence to help them come up with a solution.
- *The third step*: Brainstorm to develop alternatives. Coming up with more than one solution ables you to see which one can actually work.
- *The fourth step*: List pros and cons of each alternative. With the list of pros and cons, you can eliminate the solutions that have more cons than pros, making your decision easier.
- *The fifth step*: Make the decision. Once you analyse each solution, you should pick the one that has many pros and is a solution that everyone can agree with.
- *The sixth step*: Immediately take action. Once the decision is picked, you should implement it right away.
- *The seventh step*: Learn from, and reflect on the decision making. This step allows you to see what you did right and wrong when coming up, and putting the decision to use.

COGNITIVE AND PERSONAL BIASES

Biases can creep into our decision making processes. Many different people have made a decision about the same question and then craft potential cognitive interventions aimed at improving decision making outcomes.

A list of some of the more commonly debated cognitive biases:

- *Selective search for evidence*: We tend to be willing to gather facts that support certain conclusions but disregard other facts that support different conclusions. Individuals who are highly defensive in this manner show significantly greater left prefrontal cortex activity as measured by EEG than do less defensive individuals.
- *Premature termination of search for evidence*: We tend to accept the first alternative that looks like it might work.
- *Inertia*: Unwillingness to change thought patterns that we have used in the past in the face of new circumstances.

- *Selective perception*: We actively screen-out information that we do not think is important. In one demonstration of this effect, discounting of arguments with which one disagrees was decreased by selective activation of right prefrontal cortex.
- *Wishful thinking or optimism bias*: We tend to want to see things in a positive light and this can distort our perception and thinking.
- Choice-supportive bias occurs when we distort our memories of chosen and rejected options to make the chosen options seem more attractive.
- *Recency:* We tend to place more attention on more recent information and either ignore or forget more distant information. The opposite effect in the first set of data or other information is termed Primacy effect.
- *Repetition bias*: A willingness to believe what we have been told most often and by the greatest number of different sources.
- *Anchoring and adjustment*: Decisions are unduly influenced by initial information that shapes our view of subsequent information.
- *Group think*: Peer pressure to conform to the opinions held by the group.
- *Source credibility bias*: We reject something if we have a bias against the person, organization, or group to which the person belongs: We are inclined to accept a statement by someone we like.
- *Incremental decision making and escalating commitment*: We look at a decision as a small step in a process and this tends to perpetuate a series of similar decisions. This can be contrasted with zero-based decision making.
- *Attribution asymmetry*: We tend to attribute our success to our abilities and talents, but we attribute our failures to bad luck and external factors. We attribute other's success to good luck, and their failures to their mistakes.
- *Role fulfillment*: We conform to the decision making expectations that others have of someone in our position.
- *Underestimating uncertainty and the illusion of control:* We tend to underestimate future uncertainty because we tend to believe we have more control over events than we really do. We believe we have control to minimize potential problems in our decisions.

BOUNDED RATIONALITY

Bounded rationality is the idea that in decision making, rationality of individuals is limited by the information they have, the cognitive limitations of their minds, and the finite amount of time they have to make decisions. It was proposed by Herbert Simon as an alternative basis for the mathematical modeling of decision making, as used in economics and related disciplines; it complements rationality as optimization, which views decision making as a fully rational process of finding an optimal choice given the information available.

Another way to look at bounded rationality is that, because decision-makers lack the ability and resources to arrive at the optimal solution, they instead apply their rationality only after having greatly simplified the choices available. Thus the decision-maker is a satisficer, one seeking a satisfactory solution rather than the optimal one.

Simon used the analogy of a pair of scissors, where one blade is the "cognitive limitations" of actual humans and the other the "structures of the environment"; minds with limited cognitive resources can thus be successful by exploiting pre-existing structure and regularity in the environment. Some models of human behaviour in the social sciences assume that humans can be reasonably approximated or described as "rational" entities.

Many economics models assume that people are on average rational, and can in large enough quantities be approximated to act according to their preferences. The concept of bounded rationality revises this assumption to account for the fact that perfectly rational decisions are often not feasible in practice due to the finite computational resources available for making them.

MODELS OF BOUNDED RATIONALITY

The term is thought to have been coined by Herbert Simon. In Models of Man, Simon points out that most people are only partly rational, and are emotional/irrational in the remaining part of their actions. In another work, he states "boundedly rational agents experience limits in formulating and solving complex problems and in processing. Simon describes a number of dimensions along which "classical" models of rationality can be made somewhat more realistic, while sticking within the vein of fairly rigorous formalization.

These include:

- Limiting what sorts of utility functions there might be.
- Recognizing the costs of gathering and processing information.
- The possibility of having a "vector" or "multi-valued" utility function.

Simon suggests that economic agents employ the use of heuristics to make decisions rather than a strict rigid rule of optimization. They do this because of the complexity of the situation, and their inability to process and compute the expected utility of every alternative action. Deliberation costs might be high and there are often other concurrent economic activities also requiring decisions.

Daniel Kahneman proposes bounded rationality as a model to overcome some of the limitations of the rational-agent models in economic literature. As decision makers have to make decisions about how and when to decide, Ariel Rubinstein proposed to model bounded rationality by explicitly specifying decision-making procedures. This puts the study of decision procedures on the research agenda. Gerd Gigerenzer argues that most decision theorists who have discussed bounded rationality have not really followed Simon's ideas about it. Rather, they have either considered how people's decisions might be made sub-optimal by the limitations of human rationality, or have constructed elaborate optimising models of how people might cope with their inability to optimize.

Gigerenzer instead proposes to examine simple alternatives to a full rationality analysis as a mechanism for decision making, and he and his colleagues have shown that such simple heuristics frequently lead to better decisions than the theoretically optimal procedure. From a computational point of view, decision procedures can be encoded in algorithms and heuristics. Edward Tsang argues that the effective rationality of an agent is determined by its computational intelligence. Everything else being equal, an agent that has better algorithms and heuristics could make "more rational" decisions than one that has poorer heuristics and algorithms.

MANAGEMENT BY OBJECTIVES

The concept of 'Management by Objectives' was first given by Peter Drucker in 1954. It can be defined as a process whereby the employees and the superiors come together to identify common goals, the employees set their goals to be achieved, the standards to be taken as the criteria for measurement of their performance and contribution and deciding the course of action to be followed. The essence of MBO is participative goal setting, choosing course of actions and decision making. An important part of the MBO is the measurement and the comparison of the employee's actual performance with the standards set. Ideally, when employees themselves have been involved with the goal setting and the choosing the course of action to be followed by them, they are more likely to fulfill their responsibilities.

THE MBO PROCESS

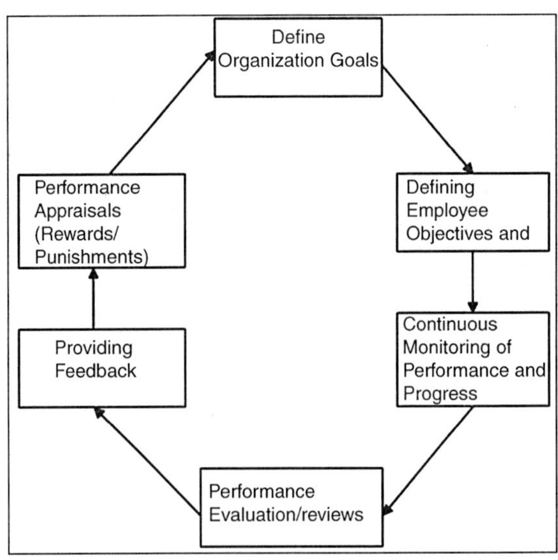

UNIQUE FEATURES AND ADVANTAGES OF MBO

The principle behind Management by Objectives is to create empowered employees who have clarity of the roles and responsibilities expected from them, understand their objectives to be achieved and thus help in the achievement of organizational as well as personal goals. Some of the important features and advantages of MBO are:

- *Clarity of goals*: With MBO, came the concept of SMART goals *i.e.* goals that are:
  - Specific
  - Measurable
  - Achievable
  - Realistic, and
  - Time bound.
- The goals thus set are clear, motivating and there is a linkage between organizational goals and performance targets of the employees.
- The focus is on future rather than on past. Goals and standards are set for the performance for the future with periodic reviews and feedback.
- *Motivation*: Involving employees in the whole process of goal setting and increasing employee empowerment increases employee job satisfaction and commitment.
- *Better communication and Coordination*: Frequent reviews and interactions between superiors and subordinates helps to maintain harmonious relationships within the enterprise and also solve many problems faced during the period.

ENVIRONMENT ANALYSIS AND DIAGNOSIS

BUSINESS ENVIRONMENT ANALYSIS

Background

Environmental analysis is a systematic process that starts from identification of environmental factors, assessing their nature and impact, auditing them to find their impact to the business, and making various profiles for positioning.

Environmental Analysis Process

A business manager should be able to analyse the environment to grasp opportunities or face the threats. Organizations need to build strength and repair their weakness available in the business environment. Therefore, this process consists not only a single steps but a process of various steps. Environmental analysis comprises scanning, monitoring, analyzing, and forecasting the business situation. Scanning is to get the relevant information from the information overload. It is to focus on the most relevant information.

Monitoring is to check the nature of the environmental factors. Analyzing requires data collection and use of different required tools and techniques. Forecasting is to find the future possibilities based on the past results and present scenario. Environmental analysis process is not static but a dynamic process. It may differ depending on the situation.

However, a general process with few common steps can be identified as the process of environmental analysis these are:

- Monitoring or identifying environmental factors,
- Scanning and selecting the relevant factors and grouping them,
- Defining variables for analysis,
- Using different methods, tools, and techniques for analysis,
- Analyzing environmental factors and forecasting,
- Designing profiles, and
- Strategic positioning and writing a report. Brief discussion is made on each of the step of this environmental analysis process.

Identifying Environmental Factors

First of all a strategist should identify all the relevant factors that might affect his or her business. In this process, one should first know what the internal areas of the business are.

This includes all the systems, internal structure, strategies followed, and culture of the organization. All these areas can be covered into the five functional areas in classical approach. Similarly, a business daily interacts with the close environmental components outside the business such as customer, competitor, and supplier. It might cover all other stakeholders such as trade union, media, and pressure group. Furthermore, general such business environment factors as political-legal, economic, sociocultural, and technological factors are to be identified

Scanning and Selecting Relevant and Key Factors

Out of all the business environmental factors, a strategist should focus only on the relevant factors for further analysis. All the factors are not equally important and affecting to the business. In this context, a strategist has to scan the environmental trend to select only the most affecting environmental factors from the information overload. This step paves the way of environment analysis and forecasting.

Defining Variables for Analysis

Selected environmental factors are to be further specified into the variables. A concept can be interpreted into different variables. For example, political situation can be measured using few variables such as instability, reliability, and long-term effect. Economic environment might cover many variables such as Per Capita, GDP, and Economic policies that can be further classified into many other variables. Variables are the basis of measurement in environmental analysis process. Variables can be compared, grouped, correlated, and predicted to find the clearer picture of the broader concept. It is, therefore, necessary to define the variables first in any kind of analysis including the environmental analysis.

Using Different Methods, Techniques, and Tools

Different types of methods, tools, and techniques are used for analysis. Some of the major methods of analysis can be Scenario Building, Benchmarking, and Network methods.

Scenario presents overall picture of its total system with affecting factors. Benchmarking is to find the best standard in an industry and to compare the one's strengths and weakness with the standard.

Network method is to assess organizational systems and its outside environment to find the strength and weakness, opportunity and threats of an organization.... Some of the techniques of primary information collection can be Delphi, Brainstorming, Survey, and Historical enquiry.

Delphi technique collects independent information from the experts without mixing them. Brainstorming is information collection technique being open minded without criticizing others.

Survey is to design questions and to ask them to the participants whereas the historical enquiry is a kind of case analysis of past period. Analysis tools can be statistical such general descriptive tools as mean, median, mode, frequency. Tools can be inferential as ANOVA, correlation, regression, factor, cluster, and multiple regression analysis.

There are many tools of analyzing functional areas. Finance and accounting use mostly profitability, leverage, fund flow and other similar accounting and financial tools for analysis. Human resources use employee turnover, training, satisfaction and many others as the basis of evaluating strength and weakness. Production area is assessed using quality control, productivity, breakdown, and many others. Similarly, marketing effectiveness is judged from the sales volume and market coverage. Research and development is perceived successful if it can really develop the strength in an organization.

Forecasting Environmental Factors

Collecting relevant information from the selected areas and to identify the variables in such areas are the basics of analysis. Analyzing the past information to predict the future is the main objective of this step. As discussed earlier, use of different methods, techniques, and tools comes under the analysis process. It is, therefore, a comprehensive process that analyses collected information using different tools and techniques.

Designing Profiles

After analyzing the environmental factors they are recorded into the profiles. Such profiles record each component or variables into left side and their positive, negative, or neutral indicators including their statement in the right side. Internal areas are recorded in Strategic Advantages Profile and external areas are recorded in Environmental Threat and Opportunity Profile. Strength, Weakness, Opportunity, and Threat profile can be designed combining both of these two profiles into one. There are varieties of reporting formats or profiles used for external and internal business environment analysis. Environmental Threat and Opportunity Profile is commonly used to report the external environmental situation whereas Strategic Advantages Profile to report the internal environmental situation.

Both of these profiles can be merged into Strength- Weakness-Opportunity-Threat profile. David used External Factor Evaluation Matrix to present weighted score of external environmental factors. Similarly, he used Internal Factor Evaluation Matrix to make the reporting of internal environmental audit. Whellen and Hunger used External Factors Analysis Summary and Internal Factors Analysis Summary that are presented in annex-.... Environmental threats and opportunities profile is a commonly used profile related to external business environment. Strategic advantages profile is related to internal business environment. Nowadays, strength and weakness and opportunities and threats profile has become very popular. Present writing pursued the approach of reporting external and internal business environment using the same approach.

Preparing ETOP

Environmental threat and opportunity profile is referred as ETOP profile. It identifies the relevant environmental factors. Such factors might be general environmental factors and task environment factors. Thereafter, it is necessary to identify their nature. Some factors are positive to the organization whereas others are negative. Therefore, it is necessary to find out their impact to the organization. Positive, neutral, and negative sign in ETOP denotes the relevant impact of environmental factors.

Preparing SAP

Strategic advantage profile is known as SAP. It shows strength and weakness of an organization. Preparation of SAP is very similar process to the ETOP. There are generally five functional areas in most of the organizations. These areas are Production or Operation, Finance or Accounting, Marketing or Distribution, Human Resource and Corporate Planning, and Research and Development. These functional areas are listed to identify their relative strength and weakness in SAP. Very similar to the

ETOP, positive, neutral, and negative signs are denoted and brief description is written in SAP profile. Each functional area is very broad having many components inside. All these described profiles provide a clear picture to understand the strategic position of an organization.

Strategic Position and Report Writing

After analysis of business environment a strategist knows the actual situation and can make some future forecasting based on the environmental analysis. After preparing the profiles strategists prepare formal report that describes the business environment. The report might present issues and best strengths of business environment in a systematic process. One can draw future strategies based on the strategic analysis followed. In conclusion, a strategist or a manager first identifies the relevant environmental factors then analyses using different tools and techniques to find out the actual situation. This overall process is sometimes known as SWOT analysis, environmental scanning, environmental analysis, or monitoring-forecasting. This process is very important for a manager to make his or her organization success by choosing the best available alternative strategy.

Some Approaches of Specific Environmental Scanning

A brief discussion is made on the potential approaches, tools, and techniques that are commonly used to analyse the sector wise business environment. The sectors of business environment are external business environment, industry level business environment, and internal business environment.

DIAGNOSIS

Diagnosis is the identification of the nature and cause of anything. Diagnosis is used in many different disciplines with variations in the use of logics, analytics, and experience to determine the cause and effect relationships. In systems engineering and computer science, diagnosis is typically used to determine the causes of symptoms, mitigations for problems, and solutions to issues.

STRATEGY FORMULATION

It is useful to consider strategy formulation as part of a strategic management process that comprises three phases: diagnosis, formulation, and implementation. Strategic management is an ongoing process to develop and revise future-oriented strategies that allow an organization to achieve its objectives, considering its capabilities, constraints, and the environment in which it operates.

Diagnosis includes:

- Performing a situation analysis, including identification and evaluation of current mission, strategic objectives, strategies, and results, plus major strengths and weaknesses;

- Analyzing the organization's external environment, including major opportunities and threats; and
- Identifying the major critical issues, which are a small set, typically two to five, of major problems, threats, weaknesses, and/or opportunities that require particularly high priority attention by management.

Formulation, the second phase in the strategic management process, produces a clear set of recommendations, with supporting justification, that revise as necessary the mission and objectives of the organization, and supply the strategies for accomplishing them. In formulation, we are trying to modify the current objectives and strategies in ways to make the organization more successful.

This includes trying to create "sustainable" competitive advantages—although most competitive advantages are eroded steadily by the efforts of competitors.

A good recommendation should be: effective in solving the stated problem, practical feasible within a reasonable time frame, cost-effective, not overly disruptive, and acceptable to key "stakeholders" in the organization. It is important to consider "fits" between resources plus competencies with opportunities, and also fits between risks and expectations.

There are four primary steps in this phase:

1. Reviewing the current key objectives and strategies of the organization, which usually would have been identified and evaluated as part of the diagnosis
2. Identifying a rich range of strategic alternatives to address the three levels of strategy formulation outlined below, including but not limited to dealing with the critical issues
3. Doing a balanced evaluation of advantages and disadvantages of the alternatives relative to their feasibility plus expected effects on the issues and contributions to the success of the organization
4. Deciding on the alternatives that should be implemented or recommended.

In organizations, and in the practice of strategic management, strategies must be implemented to achieve the intended results. The most wonderful strategy in the history of the world is useless if not implemented successfully. This third and final stage in the strategic management process involves developing an implementation plan and then doing whatever it takes to make the new strategy operational and effective in achieving the organization's objectives. The remainder of this stage focuses on strategy formulation, and is organized into six parts:

- Three Aspects of Strategy Formulation, Corporate-Level Strategy, Competitive Strategy, Functional Strategy, Choosing Strategies, and Troublesome Strategies.

THREE ASPECTS OF STRATEGY FORMULATION

The following three aspects or levels of strategy formulation, each with a different focus, need to be dealt with in the formulation phase of strategic management. The three sets of recommendations must be internally consistent and fit together in a mutually supportive manner that forms an integrated hierarchy of strategy, in the order given.

Corporate Level Strategy

In this aspect of strategy, we are concerned with broad decisions about the total organization's scope and direction. Basically, we consider what changes should be made in our growth objective and strategy for achieving it, the lines of business we are in, and how these lines of business fit together.

It is useful to think of three components of corporate level strategy:

1. Growth or directional strategy,
2. Portfolio strategy and
3. Parenting strategy.

Competitive Strategy (Often Called Business Level Strategy)

This involves deciding how the company will compete within each line of business (LOB) or strategic business unit (SBU).

Functional Strategy

These more localized and shorter-horizon strategies deal with how each functional area and unit will carry out its functional activities to be effective and maximize resource productivity.

CORPORATE LEVEL STRATEGY

This comprises the overall strategy elements for the corporation as a whole, the grand strategy, if you please.

Corporate strategy involves four kinds of initiatives:

- Making the necessary moves to establish positions in different businesses and achieve an appropriate amount and kind of diversification. A key part of corporate strategy is making decisions on how many, what types, and which specific lines of business the company should be in. This may involve deciding to increase or decrease the amount and breadth of diversification. It may involve closing out some LOB's, adding others, and/or changing emphasis among LOB's.
- Initiating actions to boost the combined performance of the businesses the company has diversified into: This may involve vigorously pursuing rapid-growth strategies in the most promising LOB's, keeping the other core businesses healthy, initiating turnaround

efforts in weak-performing LOB's with promise, and dropping LOB's that are no longer attractive or don't fit into the corporation's overall plans. It also may involve supplying financial, managerial, and other resources, or acquiring and/or merging other companies with an existing LOB.

- Pursuing ways to capture valuable cross-business strategic fits and turn them into competitive advantages—especially transferring and sharing related technology, procurement leverage, operating facilities, distribution channels, and/or customers.
- Establishing investment priorities and moving more corporate resources into the most attractive LOB's.

It is useful to organize the corporate level strategy considerations and initiatives into a framework with the following three main strategy components: growth, portfolio, and parenting.

What Should be Our Growth Objective and Strategies

Growth objectives can range from drastic retrenchment through aggressive growth. Organizational leaders need to revisit and make decisions about the growth objectives and the fundamental strategies the organization will use to achieve them.

There are forces that tend to push top decision-makers towards a growth stance even when a company is in trouble and should not be trying to grow, for example bonuses, stock options, fame, ego. Leaders need to resist such temptations and select a growth strategy stance that is appropriate for the organization and its situation. Stability and retrenchment strategies are underutilized. Some of the major strategic alternatives for each of the primary growth stances.

Growth Strategies

All growth strategies can be classified into one of two fundamental categories: concentration within existing industries or diversification into other lines of business or industries. When a company's current industries are attractive, have good growth potential, and do not face serious threats, concentrating resources in the existing industries makes good sense. Diversification tends to have greater risks, but is an appropriate option when a company's current industries have little growth potential or are unattractive in other ways. When an industry consolidates and becomes mature, unless there are other markets to seek, a company may have no choice for growth but diversification.

There are two basic concentration strategies, vertical integration and horizontal growth. Diversification strategies can be divided into related and unrelated diversification. Each of the resulting four core categories of strategy alternatives can be achieved internally through investment and development, or externally through mergers, acquisitions, and/or strategic alliances—thus producing eight major growth strategy categories.

- *Vertical Integration*: This type of strategy can be a good one if the company has a strong competitive position in a growing, attractive industry. A company can grow by taking over functions earlier in the value chain that were previously provided by suppliers or other organizations. This strategy can have advantages, *e.g.*, in cost, stability and quality of components, and making operations more difficult for competitors. However, it also reduces flexibility, raises exit barriers for the company to leave that industry, and prevents the company from seeking the best and latest components from suppliers competing for their business. A company also can grow by taking over functions forward in the value chain previously provided by final manufacturers, distributors, or retailers. This strategy provides more control over such things as final products/services and distribution, but may involve new critical success factors that the parent company may not be able to master and deliver. For example, being a world-class manufacturer does not make a company an effective retailer. Some writers claim that backward integration is usually more profitable than forward integration, although this does not have general support. In any case, many companies have moved towards less vertical integration during the last decade or so, replacing significant amounts of previous vertical integration with outsourcing and various forms of strategic alliances.
- *Horizontal Growth*: This strategy alternative category involves expanding the company's existing products into other locations and/or market segments, or increasing the range of products/services offered to current markets, or a combination of both. It amounts to expanding sideways at the point in the value chain that the company is currently engaged in. One of the primary advantages of this alternative is being able to choose from a fairly continuous range of choices, from modest extensions of present products/markets to major expansions—each with corresponding amounts of cost and risk.
- *Related Diversification*: In this alternative, a company expands into a related industry, one having synergy with the company's existing lines of business, creating a situation in which the existing and new lines of business share and gain special advantages from commonalities such as technology, customers, distribution, location, product or manufacturing similarities, and government access. This is often an appropriate corporate strategy when a company has a strong

competitive position and distinctive competencies, but its existing industry is not very attractive.

- *Unrelated Diversification (aka Conglomerate Diversification)*: This fourth major category of corporate strategy alternatives for growth involves diversifying into a line of business unrelated to the current ones. The reasons to consider this alternative are primarily seeking more attractive opportunities for growth in which to invest available funds, risk reduction, and/or preparing to exit an existing line of business. Further, this may be an appropriate strategy when, not only the present industry is unattractive, but the company lacks outstanding competencies that it could transfer to related products or industries. However, because it is difficult to manage and excel in unrelated business units, it can be difficult to realise the hoped-for value added.
- *Mergers, Acquisitions, and Strategic Alliances*: Each of the four growth strategy categories just discussed can be carried out internally or externally, through mergers, acquisitions, and/or strategic alliances. Of course, there also can be a mixture of internal and external actions. Various forms of strategic alliances, mergers, and acquisitions have emerged and are used extensively in many industries today. They are used particularly to bridge resource and technology gaps, and to obtain expertise and market positions more quickly than could be done through internal development. They are particularly necessary and potentially useful when a company wishes to enter a new industry, new markets, and/or new parts of the world. Despite their extensive use, a large share of alliances, mergers, and acquisitions fall far short of expected benefits or are outright failures.

For example, one study published in Business Week in 1999 found that 61 per cent of alliances were either outright failures or "limping along." Research on mergers and acquisitions includes a Mercer Management Consulting study of all mergers from 1990 to 1996 which found that nearly half "destroyed" shareholder value; an A. T. Kearney study of 115 multibillion-dollar, global mergers between 1993 and 1996 where 58 per cent failed to create "substantial returns for shareholders" in the form of dividends and stock price appreciation; and a Price-Waterhouse-Coopers study of 97 acquisitions over $500 million from 1994 to 1997 in which two-thirds of the buyer's stocks dropped on announcement of the transaction and a third of these were still lagging a year later.

Many reasons for the problematic record have been cited, including paying too much, unrealistic expectations, inadequate due diligence, and conflicting corporate cultures; however, the most powerful contributor to success or failure is inadequate attention to the merger integration process. Although the lawyers and investment bankers may consider a deal done when the papers are signed and they receive their fees, this should be merely an incident in a multi-year process of integration that began before the signing and continues far beyond.

Stability Strategies

There are a number of circumstances in which the most appropriate growth stance for a company is stability, rather than growth. Often, this may be used for a relatively short period, after which further growth is planned. Such circumstances usually involve a reasonable successful company, combined with circumstances that either permit a period of comfortable coasting or suggest a pause or caution.

In which the actual strategy actions are similar, but differing primarily in the circumstances motivating the choice of a stability strategy and in the intentions for future strategic actions.

- *Pause and Then Proceed*: This stability strategy alternative may be appropriate in either of two situations:
  - The need for an opportunity to rest, digest, and consolidate after growth or some turbulent events - before continuing a growth strategy, or
  - An uncertain or hostile environment in which it is prudent to stay in a "holding pattern" until there is change in or more clarity about the future in the environment.
- *No Change*: This alternative could be a cop-out, representing indecision or timidity in making a choice for change. Alternatively, it may be a comfortable, even long-term strategy in a mature, rather stable environment, *e.g.*, a small business in a small town with few competitors.
- *Grab Profits While You Can*: This is a non-recommended strategy to try to mask a deteriorating situation by artificially supporting profits or their appearance, or otherwise trying to act as though the problems will go away. It is an unstable, temporary strategy in a worsening situation, usually chosen either to try to delay letting stakeholders know how bad things are or to extract personal gain before things collapse. Recent terrible examples in the USA are Enron and WorldCom.

Retrenchment Strategies

- *Turnaround*: This strategy, dealing with a company in serious trouble, attempts to resuscitate or revive the company through a combination of contraction and consolidation. Although difficult, when done very effectively it can succeed in both retaining enough key employees and revitalizing the company.

- *Captive Company Strategy*: This strategy involves giving up independence in exchange for some security by becoming another company's sole supplier, distributor, or a dependent subsidiary.
- *Sell Out*: If a company in a weak position is unable or unlikely to succeed with a turnaround or captive company strategy, it has few choices other than to try to find a buyer and sell itself.
- *Liquidation*: When a company has been unsuccessful in or has none of the previous three strategic alternatives available, the only remaining alternative is liquidation, often involving a bankruptcy. There is a modest advantage of a voluntary liquidation over bankruptcy in that the board and top management make the decisions rather than turning them over to a court, which often ignores stockholders' interests.

What Should Be Our Portfolio Strategy

This second component of corporate level strategy is concerned with making decisions about the portfolio of lines of business or strategic business units not the company's portfolio of individual products. Portfolio matrix models can be useful in re-examining a company's present portfolio. The purpose of all portfolio matrix models is to help a company understand and consider changes in its portfolio of businesses, and also to think about allocation of resources among the different business elements.

The two primary models are the BCG Growth-Share Matrix and the GE Business Screen. These models consider and display on a two-dimensional graph each major SBU in terms of some measure of its industry attractiveness and its relative competitive strength.

The BCG Growth-Share Matrix model considers two relatively simple variables: growth rate of the industry as an indication of industry attractiveness, and relative market share as an indication of its relative competitive strength.

The GE Business Screen, also associated with McKinsey, considers two composite variables, which can be customised by the user, for industry attractiveness and competitive strength The best test of the business portfolio's overall attractiveness is whether the combined growth and profitability of the businesses in the portfolio will allow the company to attain its performance objectives.

Related to this overall criterion are such questions as:

- Does the portfolio contain enough businesses in attractive industries?
- Does it contain too many marginal businesses or question marks?
- Is the proportion of mature/declining businesses so great that growth will be sluggish?
- Are there some businesses that are not really needed or should be divested?

- Does the company have its share of industry leaders, or is it burdened with too many businesses in modest competitive positions?
- Is the portfolio of SBU's and its relative risk/growth potential consistent with the strategic goals?
- Do the core businesses generate dependable profits and/or cash flow?
- Are there enough cash-producing businesses to finance those needing cash
- Is the portfolio overly vulnerable to seasonal or recessionary influences?
- Does the portfolio put the corporation in good position for the future?

It is important to consider diversification vs. concentration while working on portfolio strategy, *i.e.*, how broad or narrow should be the scope of the company. It is not always desirable to have a broad scope. Single-business strategies can be very successful.

Some of the advantages of a narrow scope of business are:

- Less ambiguity about who we are and what we do;
- Concentrates the efforts of the total organization, rather than stretching them across many lines of business;
- Through extensive hands-on experience, the company is more likely to develop distinctive competence; and
- Focuses on long-term profits.

However, having a single business puts "all the eggs in one basket," which is dangerous when the industry and/or technology may change. Diversification becomes more important when market growth rate slows. Building stable shareholder value is the ultimate justification for diversifying—or any strategy.

What Should be O– ur Parenting Strategy?

This third component of corporate level strategy, relevant for a multi-business company is concerned with how to allocate resources and manage capabilities and activities across the portfolio of businesses.

It includes evaluating and making decisions on the following:

- Priorities in allocating resources
- What are critical success factors in each business unit, and how can the company do well on them
- Coordination of activities and transfer of capabilities among business units
- How much integration of business units is desirable.

COMPETITIVE (BUSINESS LEVEL) STRATEGY

In this second aspect of a company's strategy, the focus is on how to compete successfully in each of the lines of business the company has chosen to engage in.

The central thrust is how to build and improve the company's competitive position for each of its lines of business. A company has competitive advantage whenever it can attract customers and defend against competitive forces better than its rivals. Companies want to develop competitive advantages that have some sustainability.

Successful competitive strategies usually involve building uniquely strong or distinctive competencies in one or several areas crucial to success and using them to maintain a competitive edge over rivals.

Some examples of distinctive competencies are superior technology and/or product features, better manufacturing technology and skills, superior sales and distribution capabilities, and better customer service and convenience.

- Competitive strategy is about being different. It means deliberately choosing to perform activities differently or to perform different activities than rivals to deliver a unique mix of value.
- The essence of strategy lies in creating tomorrow's competitive advantages faster than competitors mimic the ones you possess today.

We will consider competitive strategy by using Porter's four generic strategies as the fundamental choices, and then adding various competitive tactics.

PORTER'S FOUR GENERIC COMPETITIVE STRATEGIES

He argues that a business needs to make two fundamental decisions in establishing its competitive advantage:

- Whether to compete primarily on price or to compete through providing some distinctive points of differentiation that justify higher prices, and
- How broad a market target it will aim at. These two choices define the following four generic competitive strategies. which he argues cover the fundamental range of choices.

Overall Price (Cost) Leadership

Overall Price (Cost) Leadership appealing to a broad cross-section of the market by providing products or services at the lowest price. This requires being the overall low-cost provider of the products or services. Implementing this strategy successfully requires continual, exceptional efforts to reduce costs—without excluding product features and services that buyers consider essential. It also requires achieving cost advantages in ways that are hard for competitors to copy or match.

Some conditions that tend to make this strategy an attractive choice are:

- The industry's product is much the same from seller to seller
- The marketplace is dominated by price competition, with highly price-sensitive buyers

- There are few ways to achieve product differentiation that have much value to buyers
- Most buyers use product in same ways—common user requirements
- Switching costs for buyers are low
- Buyers are large and have significant bargaining power

Differentiation

Differentiation appealing to a broad cross-section of the market through offering differentiating features that make customers willing to pay premium prices, *e.g.*, superior technology, quality, prestige, special features, service, convenience. Success with this type of strategy requires differentiation features that are hard or expensive for competitors to duplicate. Sustainable differentiation usually comes from advantages in core competencies, unique company resources or capabilities, and superior management of value chain activities.

Some conditions that tend to favour differentiation strategies are:

- There are multiple ways to differentiate the product/service that buyers think have substantial value
- Buyers have different needs or uses of the product/service
- Product innovations and technological change are rapid and competition emphasizes the latest product features
- Not many rivals are following a similar differentiation strategy

Price (Cost) Focus

Price (Cost) Focus a market niche strategy, concentrating on a narrow customer segment and competing with lowest prices, which, again, requires having lower cost structure than competitors.

Some conditions that tend to favour focus are:

- The business is new and/or has modest resources
- The company lacks the capability to go after a wider part of the total market
- Buyers' needs or uses of the item are diverse; there are many different niches and segments in the industry
- Buyer segments differ widely in size, growth rate, profitability, and intensity in the five competitive forces, making some segments more attractive than others
- Industry leaders don't see the niche as crucial to their own success
- Few or no other rivals are attempting to specialize in the same target segment

Differentiation Focus

Differentiation Focus a second market niche strategy, concentrating on a narrow customer segment and competing through differentiating features.

Best-Cost Provider Strategy

Best-Cost Provider Strategy This is a strategy of trying to give customers the best cost/value combination, by incorporating key good-or-better product characteristics at a lower cost than competitors. This strategy is a mixture or hybrid of low-price and differentiation, and targets a segment of value-conscious buyers that is usually larger than a market niche, but smaller than a broad market.

Successful implementation of this strategy requires the company to have the resources, skills, capabilities to incorporate up-scale features at lower cost than competitors.

This strategy could be attractive in markets that have both variety in buyer needs that make differentiation common and where large numbers of buyers are sensitive to both price and value. Porter might argue that this strategy is often temporary, and that a business should choose and achieve one of the four generic competitive strategies above.

Otherwise, the business is stuck in the middle of the competitive marketplace and will be out-performed by competitors who choose and excel in one of the fundamental strategies.

His argument is analogous to the threats to a tennis player who is standing at the service line, rather than near the baseline or getting to the net. However, others present examples of companies who seem to be able to pursue successfully a best-cost provider strategy, with stability.

COMPETITIVE TACTICS

Although a choice of one of the generic competitive strategies provides the foundation for a business strategy, there are many variations and elaborations. Among these are various tactics that may be useful.

Timing Tactics

When to make a strategic move is often as important as what move to make. We often speak of first-movers, second-movers or rapid followers, and late movers.

Each tactic can have advantages and disadvantages.

Being a first-mover can have major strategic advantages when:

- Doing so builds an important image and reputation with buyers;
- Early adoption of new technologies, different components, exclusive distribution channels, etc. can produce cost and/or other advantages over rivals;
- First-time customers remain strongly loyal in making repeat purchases; and
- Moving first makes entry and imitation by competitors hard or unlikely.

However, being a second- or late-mover isn't necessarily a disadvantage. There are cases in which the first-mover's skills, technology, and strategies are easily copied or even surpassed by later-movers, allowing them to catch or pass the first-mover in a relatively short period, while having the advantage of minimizing risks by waiting until a new market is established. Sometimes, there are advantages to being a skillful follower rather than a first-mover, *e.g.*, when:

- Being a first-mover is more costly than imitating and only modest experience curve benefits accrue to the leader
- The products of an innovator are somewhat primitive and do not live up to buyer expectations, thus allowing a clever follower to win buyers away from the leader with better performing products;
- Technology is advancing rapidly, giving fast followers the opening to leapfrog a first-mover's products with more attractive and full-featured second- and third-generation products; and
- The first-mover ignores market segments that can be picked up easily.

Market Location Tactics

These fall conveniently into offensive and defensive tactics. Offensive tactics are designed to take market share from a competitor, while defensive tactics attempt to keep a competitor from taking away some of our present market share, under the onslaught of offensive tactics by the competitor.

Some offensive tactics are:

- *Frontal Assault*: Going head-to-head with the competitor, matching each other in every way. To be successful, the attacker must have superior resources and be willing to continue longer than the company attacked.
- *Flanking Maneuver*: Attacking a part of the market where the competitor is weak. To be successful, the attacker must be patient and willing to carefully expand out of the relatively undefended market niche or else face retaliation by an established competitor.
- *Encirclement*: Usually evolving from the previous two, encirclement involves encircling and pushing over the competitor's position in terms of greater product variety and/or serving more markets. This requires a wide variety of abilities and resources necessary to attack multiple market segments.
- *Bypass Attack*: Attempting to cut the market out from under the established defender by offering a new, superior type of produce that makes the competitor's product unnecessary or undesirable.
- *Guerrilla Warfare*: Using a "hit and run" attack on a competitor, with small, intermittent assaults on different market segments. This offers the possibility for even a small firm to make some gains without seriously threatening a large, established competitor and evoking some form of retaliation.

Some Defensive Tactics are:

- *Raise Structural Barriers*: Block avenues challengers can take in mounting an offensive
- *Increase Expected Retaliation*: Signal challengers that there is threat of strong retaliation if they attack
- *Reduce Inducement for Attacks*: *e.g.*, lower profits to make things less attractive. Keeping prices very low gives a new entrant little profit incentive to enter.

The general experience is that any competitive advantage currently held will eventually be eroded by the actions of competent, resourceful competitors. Therefore, to sustain its initial advantage, a firm must use both defensive and offensive strategies, in elaborating on its basic competitive strategy.

COOPERATIVE STRATEGIES

Another group of "competitive" tactics involve cooperation among companies. These could be grouped under the heading of various types of strategic alliances, which have been discussed to some extent under Corporate Level growth strategies. These involve an agreement or alliance between two or more businesses formed to achieve strategically significant objectives that are mutually beneficial.

Some are very short-term; others are longer-term and may be the first stage of an eventual merger between the companies. Some of the reasons for strategic alliances are to: obtain/share technology, share manufacturing capabilities and facilities, share access to specific markets, reduce financial/political/market risks, and achieve other competitive advantages not otherwise available.

There could be considered a continuum of types of strategic alliances, ranging from:

- Mutual service consortiums
- Licensing arrangements,
- Joint ventures
- Value-chain partnerships.

FUNCTIONAL STRATEGIES

Functional strategies are relatively short-term activities that each functional area within a company will carry out to implement the broader, longer-term corporate level and business level strategies. Each functional area has a number of strategy choices, that interact with and must be consistent with the overall company strategies.

Three basic characteristics distinguish functional strategies from corporate level and business level strategies: shorter time horizon, greater specificity, and primary involvement of operating managers. A few examples follow of functional strategy topics for the major functional areas of marketing, finance, production/operations, research and development, and human resources management. Each area needs to deal with sourcing strategy, *i.e.*, what should be done in-house and what should be outsourced? Marketing strategy deals with product/service choices and features, pricing strategy, markets to be targeted, distribution, and promotion considerations.

Financial strategies include decisions about capital acquisition, capital allocation, dividend policy, and investment and working capital management. The production or operations functional strategies address choices about how and where the products or services will be manufactured or delivered, technology to be used, management of resources, plus purchasing and relationships with suppliers.

For firms in high-tech industries, R and D strategy may be so central that many of the decisions will be made at the business or even corporate level, for example the role of technology in the company's competitive strategy, including choices between being a technology leader or follower. However, there will remain more specific decisions that are part of R and D functional strategy, such as the relative emphasis between product and process R and D, how new technology will be obtained, and degree of centralization for R and D activities.

Human resources functional strategy includes many topics, typically recommended by the human resources department, but many requiring top management approval. Examples are job categories and descriptions; pay and benefits; recruiting, selection, and orientation; career development and training; evaluation and incentive systems; policies and discipline; and management/executive selection processes.

CHOOSING THE BEST STRATEGY ALTERNATIVES

Decision making is a complex subject, worthy of a stage or book of its own. This part can only offer a few suggestions. Among the many sources for additional information, I recommend Harrison, McCall and Kaplan, and Williams.

Here are some factors to consider when choosing among alternative strategies:

- It is important to get as clear as possible about objectives and decision criteria
- The primary answer to the previous question, and therefore a vital criterion, is that the chosen strategies must be effective in addressing the "critical issues" the company faces at this time
- They must be consistent with the mission and other strategies of the organization
- They need to be consistent with external environment factors, including realistic assessments of the competitive environment and trends
- They fit the company's product life cycle position and market attractiveness/competitive strength situation

- They must be capable of being implemented effectively and efficiently, including being realistic with respect to the company's resources
- The risks must be acceptable and in line with the potential rewards
- It is important to match strategy to the other aspects of the situation, including:
  - Size, stage, and growth rate of industry;
  - Industry characteristics, including fragmentation, importance of technology, commodity product orientation, international features; and
  - Company position
- Consider stakeholder analysis and other people-related factors
- Sometimes it is helpful to do scenario construction, *e.g.*, cases with optimistic, most likely, and pessimistic assumptions.

SOME TROUBLESOME STRATEGIES TO AVOID OR USE WITH CAUTION

- *Follow the Leader*: when the market has no more room for copycat products and look-alike competitors. Sometimes such a strategy can work fine, but not without careful consideration of the company's particular strengths and weaknesses.
- *Count On Hitting Another Home Run*: *e.g.*, Polaroid tried to follow its early success with instant photography by developing "Polavision" during the mid-1970s. Unfortunately, this very expensive, instant developing, 8mm, black and white, silent motion picture camera and film was displayed at a stockholders' meeting about the time that the first beta-format video recorder was released by Sony. Polaroid reportedly wrote off at least $500 million on this venture without selling a single camera.
- *Try to Do Everything*: Establishing many weak market positions instead of a few strong ones
- *Arms Race*: Attacking the market leaders head-on without having either a good competitive advantage or adequate financial strength; making such aggressive attempts to take market share that rivals are provoked into strong retaliation and a costly "arms race." Such battles seldom produce a substantial change in market shares; usual outcome is higher costs and profitless sales growth
- *Put More Money On a Losing Hand*: one version of this is allocating R and D efforts to weak products instead of strong products
- *Over-optimistic Expansion*: Using high debt to finance investments in new facilities and equipment, then getting trapped with high fixed costs when demand turns down, excess capacity appears, and cash flows are tight

- *Unrealistic Status-Climbing*: Going after the high end of the market without having the reputation to attract buyers looking for name-brand, prestige goods
- *Selling the Sizzle Without the Steak*: Spending more money on marketing and sales promotions to try to get around problems with product quality and performance. Depending on cosmetic product improvements to serve as a substitute for real innovation and extra customer value.

6

Improvement in Biotechnological Organisms

INTRODUCTION

In the last several decades, a group of microbial secondary metabolites, the antibiotics, has emerged as one of the most powerful tools for combating disease. So important are antibiotics as chemotherapeutic agents that much of the effort in searching for useful bioactive microbial products has been directed towards the search for them. Thousands of secondary metabolites are, however, known and they include not only antibiotics, but also pigments, toxins, pheromones, enzyme inhibitors, immunomodulating agents, receptor antagonists and agonists, pesticides, antitumor agents and growth promoters of animals and plants. When appropriate screening has been done on secondary metabolites, numerous drugs outside antibiotics have been found. Some of such nonantibiotic drugs are shown in Table.

Table. Some Microbial Metabolites with Non-antibiotic Pharmacological Activity

Compound	Activity	Producing Microorganism
Aspergillic acid	Antihypertensive	Aspergillus sp
Astromentin	Sommoth muscle relaxant	Monascus sp
Siolipn	Acceleration of fibrin clot	Streptomyces sioyaensis
Azaserine	Antidiuretic, antitumor	Streptomyces fragilis
Ovalicin	Immunosuppressive, antitumor	Pseudeurotum ovalis
Candicidin (and other polyene Macrolides)	Cholesterol lowering	Streptomyces noursei
Streptozotocin	Hyperglycemic, antitumor	Streptomyces achromogenes
Zygosporin A	Anti-inflammatory	Cephalosporium acremonium
Fusaric acid	Hypotensive	Fusarium oxysporum
Leupeptin family	Plasmin inhibitor	Bacillus sp
Pepstatin	Pepsin inhibitor	Aspergiluus niger
Oosponol	Dopamine β-hydroxlyase inhibitor	Oospora adringens
Fumagallin	Angiogenesis inhibitor	Aspergillus fumigatus

It seems reasonable from this to conclude that the exploitation of microbial secondary metabolites, useful to man outside antibiotics, has barely been touched. A special effort is made in this book to discuss method for assaying microbial metabolites for drugs outside of antibiotics.

This section will therefore discuss in brief general terms the principles involved in searching for microorganisms producing metabolites of economic importance. The genetic improvement of strains of organisms used in biotechnology, including microorganisms, plants and animals is also discussed.

SOURCES OF MICROORGANISMS

LITERATURE SEARCH AND CULTURE COLLECTION SUPPLY

If one was starting from scratch and had no idea which organism produced a desired industrial material, then perhaps a search in the literature, including patent literature, accompanied by contact with one or more of the established culture collections and the regulatory offices dealing with patents may provide information on potentially useful microbial cultures.

The cultures may, however, be tied to patents, and fees may be involved before the organisms are supplied, along with the right to use the patented process for producing the material. Generally, cultures are supplied for a small fee from most culture collections irrespective of whether or not the organism is part of a process patent.

METABOLITES OF ECONOMIC IMPORTANCE

Although the well-known ubiquity of microorganism implies that almost any natural ecological entity–water, air, leaves, tree trunks–may provide microorganisms, the soil is the preferred source for isolating organisms, because it is a vast reservoir of diverse organisms. Indeed microorganisms capable of utilizing virtually any carbon source will be found in soil if adequate screening methods are used.

In recent times, other 'new' habitats, especially the marine environment, have been included in habitats to be studied in searches for bioactive microbial metabolites or 'bio-mining'.

Enrichment with the Substrate utilized by the Organism being Sought

If the organism being sought is one which utilizes a particular substrate, then soil is incubated with that substrate for a period of time. The conditions of the incubation can also be used to select a specific organism. Thus, if a thermophilic organism attacking the substrate is required, then the soil is incubated at an elevated temperature. After a period of incubation, a dilution of the incubated soil is plated on a medium containing the substrate and incubated at the previous temperature (*i.e.*, elevated for thermopile search).

Organisms can then be picked out especially if some means has been devised to select them. Selection could, for instance, be based on the ability to cause clear zones in an agar plate as a result of the dissolution of particles of the substrate in the agar. In the search for _-amylase producers, the soil may be enriched with starch and subsequently suitable soil dilutions are plated on agar containing starch as the sole carbon source. Clear halos form around starch-splitting colonies against a blue background when iodine is introduced in the plate.

Continuous culture methods are a particularly convenient means of enriching for organisms from a natural source. The constant flow of nutrients over material from a natural habitat such as soil will encourage, and after a time, select for organisms able to utilize the substrate in the nutrient solution. Conditions such as pH, temperature, etc., may also be adjusted to select the organisms which will utilize the desired substrate under the given conditions. Agar platings of the outflow from the continuous culture setup are made at regular intervals to determine when an optimum population of the desired organism has developed.

Enrichment with Toxic Analogues of the Substrate Utilized by the Organism being Sought

Toxic analogues of the material where utilization is being sought may be used for enrichment, and incubated with soil. The toxic analogue will kill many organisms which utilize it. The surviving organisms are then grown on the medium with the non-toxic substrate. Under the new conditions of growth many organisms surviving from exposure to toxic analogues over-produce the desired end-products.

Testing Microbial Metabolites for Bioactive Activity

Testing for Anti-microbial activity

For the isolation of antibiotic producing organisms the metabolites of the test organism are tested for anti-microbial activity against test organisms. One of the commonest starting point is to place a soil suspension or soil particles on agar seeded with the test organism(s). Colonies around which cleared zones occur are isolated, purified, and further studied.

Testing for Enzyme Inhibition

Microorganisms whose broth cultures are able to inhibit enzymes associated with certain disease may be isolated and tested for the ability to produce drugs for combating the disease. Enzyme inhibition may be determined using one of the two methods among those discussed by Umezawa in 1982. In the first method the product of the reaction between an enzyme and its substrate is measured using spectroscopic methods.

The quantity of the inhibitor in the test sample is obtained by measurin:

- The product in the reaction mixture without the inhibitor and
- The product in the mixture with the inhibitor (*i.e.*, a broth or suitable fraction of the broth whose inhibitory potency is being tested).

The percentage inhibition (if any) is calculated by the formula:

$$\frac{(a-b)}{a}\times 100$$

The second method determines the quantity of the unreacted substrate. *For this determination the following measurements of the substrate are made*:

- With the enzyme and without the inhibitor (*i.e.*, broth being tested);
- With the enzyme and with the inhibitor and;
- Without the enzyme and without the inhibitor.

Percentage inhibition (if any) is determined by (c-a)–(c-b) x 100.The results obtained enable the assessment of the existence of enzyme inhibitors and facilitate the comparison of the inhibitory ability of broths from several sources.

Testing for Morphological Changes in Fungal Test Organisms

The effect on spore germination or change in hyphal morphology may be used to detect the presence of pharmacologically active products in the broth of a test organism. This method does not rely on the death or inhibition of microbial growth, which has been so widely used for detecting antibiotic presence in broths.

Conducting Animal Tests on the Microbial Metabolites

The effect of broth on various animal body activities such as blood pressure, immunosuppressive action, anti-coagulant activity are carried out in animals to determine the content of potentially useful drugs in the broth.

STRAIN IMPROVEMENT

Several options are open to an industrial microbiology organization seeking to maximize its profits in the face of its competitors' race for the same market. The organization may undertake more aggressive marketing tactics, including more attractive packaging while leaving its technical procedures unchanged. It may use its human resources more efficiently and hence reduce costs, or it may adopt a more efficient extraction system for obtaining the material from the fermentation broth.

The operations in the fermentor may also be improved by its use of a more productive medium, better environmental conditions, better engineering control of the fermentor processes, or it may genetically improve the productivity of the microbial strain it is using.

Strain improvement appears to be the one single factor with the greatest potential for contributing to greater profitability. While realizing the importance of strain improvement, it must be borne in mind that an improved strain could bring with it previously non-existent problems.

For example, a more highly yielding strain may require greater aeration or need more intensive foam control; the products may pose new extraction challenges, or may even require an entirely new fermentation medium. The use of a more productive strain must therefore be weighed against possible increased costs resulting from higher investments in extraction, richer media, more expensive fermentor operations and other hitherto nonexistent problems.

This possibility not withstanding, strain improvement is usually part of the programme of an industrial microbiology organization.

To appreciate the basis of strain improvement it is important to remember that the ability of any organism to make any particular product is predicated on its capability for the secretion of a particular set of enzymes. The production of the enzymes, themselves depends ultimately on the genetic make-up of the organisms.

Improvement of Strains can therefore be put down in simple term as follows:

- Regulating the activity of the enzymes secreted by the organisms;
- In the case of metabolites secreted extracellularly, increasing the permeability of the organism so that the microbial products can find these way more easily outside the cell;
- Selecting suitable producing strains from a natural population;
- Manipulation of the existing genetic apparatus in a producing organism;
- Introducing new genetic properties into the organism by recombinant DNA technology or genetic engineering.

SELECTION FROM NATURALLY OCCURRING VARIANTS

In selection of this type, naturally occurring variants which over-produce the desired product are sought. Strains which were encountered but not selected should not be automatically discarded; the better ones are usually kept as stock cultures in the organization's culture collection for possible use in future genetic manipulations. Selection from natural variants is a regular feature of industrial microbiology and biotechnology. For example, in the early days of antibiotic production the initial increase in yield was obtained in both penicillin and griseofulvin by natural variants producing higher yields in submerged rather than in surface culture.

Another example is lager beer manufacture where the constant selection of yeasts that flocculate eventually gave rise to strains which are now used for the production of the beverage. Similarly in wine fermentation yeasts were repeatedly taken from the best vats until yeasts of suitable properties were obtained. Selection of this type is not only slow but its course is largely outside the control of the biotechnologist, an intolerable condition in the highly competitive world of modern industry.

GENOME OF INDUSTRIAL ORGANISMS IN

The Manipulation of the genome for increased productivity may be done in one of two general procedures as shown in Table:

1. Manipulations not involving foreign DNA;
2. Manipulations involving foreign DNA.

Table. Methods of Manipulating the Genetic Apparatus of Industrial Organisms

A. Methods not involving foreign DNA
1.Conventional mutation
B. Methods involving DNA foreign to the organism (*i.e.* recombination)
2. Transduction
3. Conjugation
4. Transformation
5. Heterokaryosis
6. Protoplast fusion
7. Genetic engineering
8. Metabolic engineering
9. Site-directed mutation

Genome Manipulations not Involving Foreign DNA or Bases: Conventional Mutation

Nature of Conventional Mutation

The properties of any microorganism depend on the sequence of the four nucleic acid bases on its genome: adenine (A), thymine (T), cytosine (C), and guanine (G). The arrangement of these DNA bases dictates the distribution of genes and hence the nature of proteins synthesized. A mutation can therefore be described as a change in the sequence of the bases in DNA (or RNA, in RNA viruses).

It is clear that since it is the sequence of these bases which is responsible for the type of proteins (and hence enzymes) synthesized, any change in the sequence will lead ultimately to a change in the properties of the organism. Mutations occur spontaneously at a low rate in a population of microorganisms. It is this low rate of mutations which is partly responsible for the variation found in natural populations. An increased rate can however be induced by mutagens, (or mutagenic agents) which can either be physical or chemical.

Physical Agents

Ionizing Radiations

X-rays, gamma rays, alpha-particles and fast neutrons are ionizing radiations and have all been successfully used to induce mutation. X-rays are produced by commercially available machines as well as van de Graaf generators.

Gamma rays are emitted by the decay of radioactive materials such as Cobalt $_{60}$.Fast neutrons are produced by a cyclotron or an atomic pile.

Ionizing radiations are so called because they knock off the outer electrons in the atoms of biological materials (including DNA) thereby causing ionization in the molecules of DNA. As a result, highly reactive radicals are produced and these cause changes in the DNA. Some authors do not advise the use of ionizing radiations unless all other methods fail. This is party because the equipment is expensive and hence not always readily available, but also because ionizing radiations are apt to cause breakage in chromosomes.

Ultraviolet Light

The mutagenic range of ultraviolet light lies between wave length 200 and 300 nm. 'Low pressure' UV lamps used for mutagenesis emit most of their rays in the 254 nm region. The suspension of cells or spores to be mutagenized is placed in a Petri dish 2-3 cm below a 15 watt lamp and stirred either by a rocking mechanism or by a magnetic stirrer. The organisms are exposed for varying periods lasting from about 300 seconds to about 20 minutes depending on the sensitivity of the organisms. Since UV damage can be repaired by exposure to light in a process known as photo-reactivation all manipulations should be conducted under a special light source such as 25 watt yellow or red bulbs. A proportion of the organisms ranging from about 60–99.9 per cent should be killed by the radiation. The preference of workers as to the amount of kill varies, but the higher the kill the more the likelihood of producing desirable mutants. Furthermore, the higher the kill, the less likely it is that the killing is due to overheating consequent on having the organism too close to the lamp. The initial concentration of the organisms should also be in the order of 107 per ml.

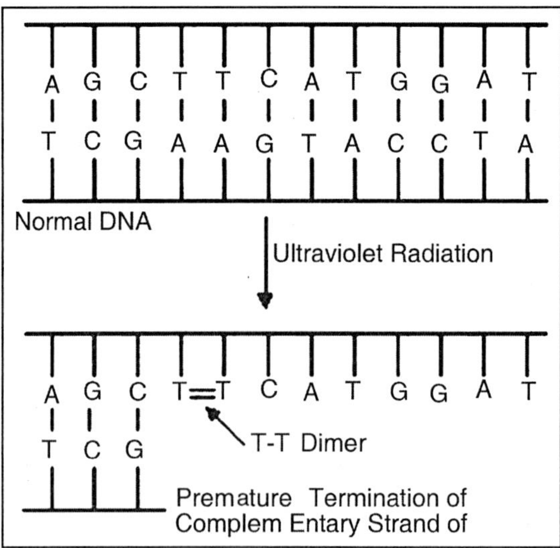

Fig. Schematic Representation of Thymine Dimerization by UV Light on DNA

The main effect of ultraviolet light on DNA is the formation of covalent bonds between adjacent pyrimidine (thymine and cytosine) bases. Thymine is mainly affected, and hence the major effect of UV light is thymine dimerization, although it can also cause thymine-cytosin and cytosin-cytosin dimers. Dimerization causes a distortion of the DNA double strand and the ultimate effect is to inhibit transcription and finally the organism dies.

Chemical Mutagens

These may be divided into three groups:

1. Those that act on DNA of resting or non-dividing organisms;
2. DNA analogues which may be incorporated into DNA during replication;
3. Those that cause frame-shift mutations.

- *Chemicals acting on resting DNA*: Some chemical mutagens, such as nitrous acid and nitrosoguanidine work by causing chemical modifications of purine and pyrimidine bases that alter their hydrogenbonding properties. For example, nitrous acid converts cytosine to uracil which then forms hydrogen bonds with adenine rather than guanine. These chemicals act on the non-dividing cell and include nitrous acid, alkylating agents and nitrosoguanidine (NTG) (also known as MNNG).
  - *Nitrous acid*: This acid is rather harmless and the mutation can be easily performed by adding 0.1 to 0.2 M of sodium nitrate to a suspension of the cells in an acid medium for various times. The acid is neutralized after suitable intervals by the addition of appropriate amounts of sodium hydroxide. The cells are plated out subsequently.
  - *Alkylating agents*: These are compounds with one or more alkyl groups which can be transferred to DNA or other molecules. Many of them are known but the following have been routinely used as mutagens: EMS (ethyl methane sulphonate), EES (ethyl ethane sulphonate) and DES (Diethyl sulphonate). They are liquids and easy to handle. Cells are treated in solutions of about 1 per cent concentration and allowed to react from ¼ hour to ½ hour and thereafter are plated out. Experimentation has to be done to decide the amount of kill that will provide a suitable amount of mutation. While some are carcinostatic (*i.e.*, stop cancers), some are carcinogenic and must be handled carefully.
  - *NTG–nitrosoguanidine*: also known as M-methyl-N-nitro-M-guanidine - MNNG: it is one of the most potent mutagens known and must therefore should be handled with care. Amounts ranging from 0.1 to 3.0 mg/ml have been used but for most mutations

the lower quantity is used. It is reported to induce mutation in closely linked genes. It is widely used in industrial microbiology.

- *Nitrogen mustards*: The most commonly used of this group of compounds is methyl-bis (Beta-chlorethyl) amine also referred to as 'HN2'. Nitrogen mustards were used for chemical warfare in World War I. Other members of the group are 'HN, ' 'HN1', or 'HN3' from the wartime code name for mustard gas, H. The number after the H denotes the number of 2-chloroethyl groups which have replaced the methyl groups in trimethylamine. A spore or cell suspension is made in HN2 (methyl-bis [Beta-chloroethyl amine]) and after exposure to various concentrations for about 30 minutes each, the reaction is ended by a decontaminating solution containing 0.7 per cent $NaHCO_3$ and 0.6 per cent glycine. The solution is then plated out for survivors. Between 0.05 and 0.1 per cent HN2 solutions in 2 per cent sodium bicarbonate solutions have been found satisfactory for Streptomyces. Sometimes the exposure time may be extended

- *Base analogues*: These are compounds which because they are similar to base nucleotides in composition may be incorporated into a dividing DNA in place of the natural base. However, this incorporation takes place only in special conditions. The best examples include 2-amino purine, a compound that resembles adenine, and 5-bromouracil (5BU), a compound that resembles thymine. The base analogs, however, do not have the hydrogen-bonding properties of the natural base. Base analogues are not useful as routine mutagens because suitable conditions for their use may be difficult to achieve. For example, with BU, incorporation occurs only when the organisms is starved of thymine.
- *Frameshift mutagens (also known as intercalating agents)*: Frameshift or intercalating agents are planar three-ringed molecules that are about the same size as a nucleotide base pair. During DNA replication, these compounds can insert or intercalate between adjacent base pairs thus pushing the nucleotides far enough apart that an extra nucleotide is often added to the growing chain during DNA replication. A mutation of this sort changes all the amino acids downstream and is very likely to create a nonfunctional product since it may differ greatly from the normal protein. Furthermore, reading frames (*i.e.*, the DNA base sequences) other than the correct one often contain stop codons which will truncate the mutant protein prematurely.

Acridines are among the best known of these mutagens, which cause a displacement or shift in the sequence of the bases. Although strongly mutagenic for some bacteriophages, acridines have not been found useful for bacteria. However, certain compounds, ICR (Institute for Cancer Research), (eg, ICR191) compounds in which an acridine nucleus is linked to an alkylating side chain, induce mutations in bacteria. Acridine, C13H9N, is an organic compound consisting of three fused benzene rings.

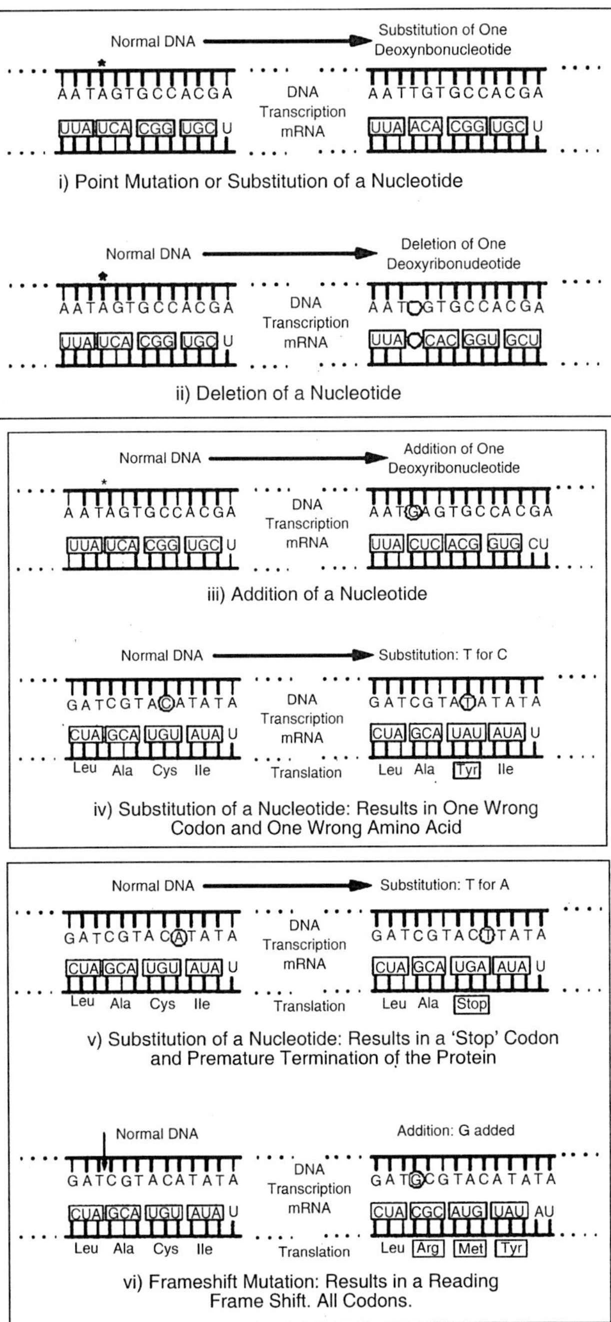

Fig. Different Types of Mutation

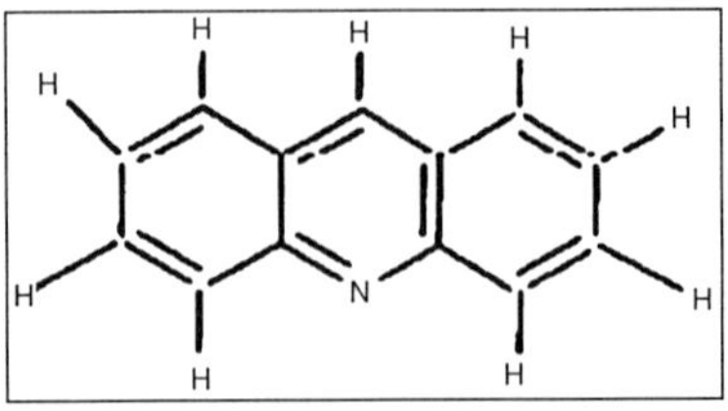

Fig. Acridine

Acridine is colourless and was first isolated from crude coal tar. It is a raw material for the production of dyes. Acridines and their derivatives are DNA and RNA binding compounds due to their intercalation abilities. Acridine Orange (3, 6- dimethylaminoacridine) is a nucleic acid selective metachromatic stain useful for cell cycle determination. Another example is ethidium bromide, which is also used as a DNA dye.

Choice of Mutagen

Mutagenic agents are numerous but not necessarily equally effective in all organisms. Should one agent fail to produce mutations then another should be tried.

Other factors besides effectiveness to be borne in mind are:

- The safety of the mutagen: many mutagens are carcinogens,
- Simplicity of technique,
- Ready availability of the necessary equipment and chemicals.

Among physical agents, UV is to be preferred since it does not require much equipment, and is relatively effective and has been widely used in industry. Chemical methods other than NTG are probably best used in combination with UV. The disadvantage of UV is that it is absorbed by glass; it is also not effective in opaque or coloured organisms.

The Practical Isolation of Mutants

There are three stages before a mutant can come into use: the organisms must be exposed to a suitable mutagen under suitable conditions; the treated cells must be exposed to conditions which ideally select for the mutant; and finally, the mutant must then be tested for productivity.

- *Exposing organisms to the mutagen*: The organism undergoing mutation should be in the haploid stage during the exposure. Bacterial cells are haploid; in fungi and actinomycetes the haploid stage is found in the spores. However, in non-sporing strains of these organisms hyphae, preferable the tips, may be used. The use of haploid is essential because many mutant genes are recessive in comparison to the parent or wild-type gene.
- *Selection for mutants*: Following exposure to the mutagen the cells should be suitably diluted and plated out to yield 50–100 colonies per

plate. The selection of mutants is greatly facilitated by relying on the morphology of the mutants or on some selectivity in-built into the medium on which the treated cells or spores are plated. When morphological mutants are selected, it is in the hope that the desired mutation is pleotropic (*i.e.*, a mutation in which change in one property is linked with a mutation in another character). The classic example of a pleotropic mutation is to be seen in the development of penicillin-yielding strains of Penicillium chrysogenum. It was found in the early days of the development work on penicillin production that after irradiation, strains of Penicillium chrysogenum with smaller colonies and which also sporulated poorly were better producers of penicillin. Similar increases of metabolite production associated with a morphological change have been observed in organisms producing other antibiotics: cycloheximide, nystatin, and tetracyclines. In citric acid production it was observed that mutants with colour in the conidia produced more of the acid; in some bacteria strains overproducing nucleic acid had a different morphological characteristic from those which did not. In-built selectivity of the medium for mutants over the parent cells may be achieved by manipulating the medium. If, for example, it is desired to select for mutants able to stand a higher concentration of alcohol, an antibiotic, or some other chemical substance, then the desired level of the material is added to the medium on which the organisms are plated. Only mutants able to survive the higher concentration will develop. Toxic analogues may also be incorporated. Mutants resisting the analogues develop and may be higher yielding than the parent.

- *Screening*: Screening must be carefully carried out with statistically organized experimentation to enable one to accept with confidence any apparent improvement in a producing organism. Shake cultures are preferred and about 6 of these of 500 ml capacity should be used. Accurate methods of identifying the desired product among a possible multitude of others should be worked out. It may also be better in industrial practice where time is important to carry out as soon as possible a series of mutations using ultraviolet, and a combination of ultraviolet and chemicals and then to test all the mutants.

Isolation of Auxotrophic Mutants

Auxotrophic mutants are those which lack the enzymes to manufacture certain required nutrients; consequently, such nutrients must therefore be added to the growth medium. In contrast the wild-type or prototrophic organisms possess all the enzymes needed to synthesize all growth requirements. As auxotrophic mutants are often used in industrial microbiology, *e.g.*, for the production of amino acids, nucleotides, etc., their production will be described briefly below. A procedure for producing auxotrophic mutants is illustrated in Fig.

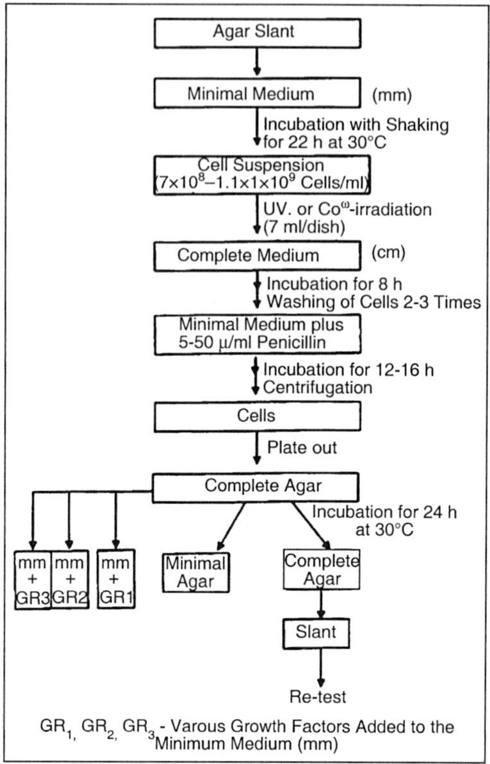

Fig. Procedure for Isolating Auxotrophic Mutants

The organism (prototroph) is transferred from a slant to a broth of the minimal medium (mm) which is the basic medium that will support the growth of the prototroph but not that of the auxotroph. The auxotroph will only grow on the complete medium, *i.e.*, the minimal medium plus the growth factor, amino-acid or vitamin which the auxotroph cannot synthesize.

The prototroph is shaken in the minimal broth for 22–24 hours, at the end of which period it is subjected to mutagenic treatment. The mutagenized cells are now grown on the complete medium for about 8 hours after which they are washed several times. The washed cells are then shaken again in minimal medium to which penicillin is added. The reason for the addition of penicillin is that the antibiotic kills only dividing cells; as only prototrophs will grow in the minimal medium these are killed off leaving the auxotrophs.

The cells are washed and plated out on the complete agar medium. In order to determine the growth factor or compound which the auxotroph cannot manufacture, an agar culture is replica-plated on to each of several plates which contain the minimal medium and various growth factors either single or mixed. The composition of the medium on which the auxotroph will grow indicates the metabolite it cannot synthesize; for example when the auxotroph requires lysine it is designated a 'lysineless' mutant.

Table. Growth of Various Mutants, Produced After Treatment of a Wild-type Organism

S/N	Complete Medium (cm)	Minimal Medium (mm)	mm + Lysine	Growth on mm + Biotin	mm + Valine	Remarks
1	+	–	–	+	–	Biotin-less mutant
2	+	–	+	–	–	Lysine-less mutant
3	+	–	–	–	+	Valine-less mutant
4	+	–	–	+	+	Biotin-and valine-less
5	+	+	+	+	+	Parent Prototype

Key:
+ = growth
– = no growth

Strain Improvement Methods Involving Foreign DNA or Bases

Transduction

Transduction is the transfer of bacterial DNA from one bacterial cell to another by means of a bacteriophage. In this process a phage attaches to, and lyses, the cell wall of its host. It then injects its DNA (or RNA) into the host. Once inside the cell the viral genome may become attached to the host DNA or remain unattached forming a plasmid.

Such a phage, which does not lyse the cell, is a temperate phage and the situation is known as lysogeny. Sometimes the viral genome may direct the host DNA to produce hundreds of copies of the phage. At the end of this manufacture the host is lysed releasing the viral particles into the medium; the new phages carry portions of the host DNA.

If one of these viral particles now invades another bacterium, but is lysogenic in the new host, the new host will acquire some nucleic acid, and hence, some properties from the previous bacterial host. This process of the acquisition of new DNA from another bacterium through a phage is transduction. Transduction is two broad types: general transduction and specialized transduction. In general transduction, host DNA from any part of the host's genetic apparatus is integrated into the virus DNA; in specialized transduction, which occurs only in some temperate phages, DNA from a specific region of the host DNA is integrated into the viral DNA and replaces some of the virus' genes. It is now possible by methods which will be discussed later under the section on genetic engineering to excise genes responsible for producing certain enzymes and attach them on the special mutant viral particles, which do not cause the lysis of their hosts.

Several hundreds of virus particles carrying the attached gene may therefore be present in one single bacterial cell following viral replication in it. The result is that the enzyme specified by the attached gene may be produced up to 1, 000-fold. Gene amplification by phage is much higher than that obtained by plasmids. The method is a well-established research tool in bacteria including actinomycetes but prospects for its use in fungi appear limited.

Transformation

Transformation is a change in genetic property of a bacterium which is brought about when foreign DNA is absorbed by, and integrates with the genome of, the donor cell. Cells in which transformation can occur are 'competent' cells. In some cases competence is artificially induced by treatment with a calcium salt. The transforming DNA must have a certain minimum length before it can be transformed. It is cut by enzymes, endonucleases, produced by the host before it is absorbed. Reports of transformation in Streptomyces spp have been made.

Transformation has been used to introduce streptomycin production into Streptomyces olivaceus with DNA from Streptomycin grisesus. Oxytetracycline producing ability was transformed into irradiated wild-type S. rimosus, using DNA from a wild-type strain. The technique has also been used to transform the production of the antifungal antibiotic thiolutin from S. pimpirin to a chlortetracycline producing S. aureofaciens which subsequently produced both antibiotics. An inactive strain of Bacillus was transformed to one producing the antibiotic bacitracin with the same method. The method has also been used to increase the level of protease and amylase production in Bacillus spp. The method therefore has good industrial potential.

Conjugation

Conjugation involves cell to cell contact or through sex pili (singular, pilus) and the transfer of plasmids. Conjugation involves a donor cell which contains a particular type of conjugative plasmid, and a recipient cell which does not. The donor strain's plasmid must possess a sex factor as a prerequisite for conjugation; only donor cells produce pili. The sex factor may on occasion transfer part of the hosts' DNA.

Mycelial 'conjugation' takes place among actinomycetes with DNA transfer as in the case of eubacteria. Among sex plasmids of actinomycetes, perhaps the two best known are plasmids SCP1 and SCP2. Plasmids play an important role in the formation of some industrial products, including many antibiotics.

Parasexual Recombination

Parasexuality is a rare form of sexual reproduction which occurs in some fungi. In parasexual recombination of nuclei in hyphae from different strains fuse, resulting in the formation of new genes. Parasexuality is important in those fungi such as Penicillium chrysogenum and Aspergiluss niger in which no sexual cycles have been observed.

It has been used to select organisms with higher yields of various industrial product such as phenoxy methyl penicillin, citric acid, and gluconic acid. Parasexuality has not become widely successful in industry because the diploid strains are unstable and tend to revert to their lower-yielding wild-type parents. More importantly is that the diploids are not always as high yielding as the parents.

Protoplast Fusion

Protoplasts are formed from bacteria, fungi, yeasts and actinomycetes when dividing cells are caused to lose their cell walls. Protoplasts may be produced in bacteria with the enzyme lysozyme, an enzyme found in tears and saliva, and capable of breaking the _-1-4 bonds linking the building blocks of the bacterial cell wall. Protoplast fusion enables recombination in strains without efficient means of conjugation such as actinomycetes. It has also been used previously to produce plant recombinants.

The technique involves the formation of stable protoplasts, fusion of protoplasts and subsequent regeneration of viable cells from the protoplasts. Fusion from mixed populations of protoplasts is greatly enhanced by the use of polyethylene glycol (PEG). Protoplast fusion has been successfully done with Bacillus subtilis and B. megaterium and among several species of Streptomyces (S. coeli-colour, S. acrimycini, S. olividans, S. pravulies) has been done between the fungi Geotrichum and Aspergillus. The method has great industrial potential and experimentally has been used to achieve higher yields of antibiotics through fusion with protoplasts from different fungi.

Site-directed Mutation

The outcome of conventional mutation which we have discussed so far, is random, the result being totally unpredictable. Recombinant DNA technology and the use of synthetic DNA now make it possible to have mutations at specific sites on the genome of the organism in a technique known as Site-Directed Mutagenesis. The mutation is caused by in vitro change directed at a specific site in a DNA molecule. The most common method involves use of a chemically synthesized oligonucleotide mutant which can hybridize with the DNA target molecule; the resulting mismatch-carrying DNA duplex may then be transfected into a bacterial cell line and the mutant strands recovered.

The DNA of the specific gene to be mutated is isolated, and the sequence of bases in the gene determined. Certain pre-determined bases are replaced and the 'new' gene is reinserted into the organism. Site-directed mutagenesis creates specific, well-defined mutations (*i.e.*, specific changes in the protein product). It has helped to raise the industrial production of enzymes, as well as to produce specific enzymes.

Metabolic Engineering

Metabolic engineering is the science which enables the rational designing or redesigning of metabolic pathways of an organism through the manipulation of the genes so as to maximize the production of biotechnological goods. In metabolic engineering, existing pathways are modified, or entirely new ones introduced through the manipulation of the genes so as to improve the yields of the microbial product, eliminate or reduce undesirable side products or shift to the production of an entirely new product. It is a modern evolution of an existing procedure which is used to induce over production of products by blocking some pathways so as to shunt productivity through another.

In the older procedure the pathways are shut off by producing mutants in which the pathways are lacking using the various mutation methods described earlier. In metabolic engineering the desired genes are isolated, modified and reintroduced into the organism. Metabolic engineering is the logical end of site-directed mutagenesis. It has been used to overproduce the amino acid isoluecine in Corynebacterium glutamicum, and ethanol by E. coli and has been employed to introduce the gene for utilizing lactose into Corynebacterium glutamicum thus making it possible for the organism to utilize whey which is plentiful and cheap.

Through metabolic engineering the gene for the utization of xylose was introduced into Klebsiella sp making it possible for the bacterium to utilize the wood sugar. possible for the bacterium to utilize the wood sugar. It is equally applicable to primary and secondary metabolites alike.

Among primary metabolites the alcohol producing adhB gene from the high alcohol yielding bacterium, Zymomonas mobilis was introduced into E. coli and Klebsiella oxytoca, enabling these organisms to produce alcohol from a wide range of sugars, hexose and pentose. Other primary metabolites which have been produced in other organisms by introducing genes from extraneous sources are carotenoids, the intermediates in the manufacture of vitamin A in the animal body, and 1, 3 propanediol (1, 3 PD) an intermediate in the synthesis of polyesters.

1, 3 PD is currently derived from petroleum and is expensive to produce. 1, 3 PD has been produced by E. coli carrying genes from Klebsiella pneumoniae able to anaerobically produce the diol. Among secondary metabolites, increase in the production of existing antibiotics, and the production of new antibiotics and anti-tumor agents have been enabled by metabolic engineering. The transfer of genes from Streptomyces erythreus to Strep lividans facilitated the production of erythromycin in the latter organism.

In the field of anti-tumor drugs, epirubicin has less cardiotoxicity than others such as the more frequently prescribed doxorubicin. The chemical production of epirubicin is complicated and requires seven steps. However using a metabolic engineering method in which the erythromycin biosynthetic gene was introduced into Strep peucetius it has been possible to produce it directly by fermentation.

Genetic Engineering

Genetic engineering, also known as recombinant DNA technology, molecular cloning or gene cloning. has been defined as the formation of new combinations of heritable material by the insertion of nucleic acid molecules produced by whatever means outside the cell, into any virus, bacterial plasmid or other vector system so as to allow their incorporation into host organisms in which they do not naturally occur but in which they are capable of continued propagation The DNA to be inserted into the host bacterium may come from a eucaryotic cell, a prokaryotic cell or may even be synthesized chemically.

The vector-foreign DNA complex which is introduced into the host DNA is sometimes known as a DNA chimera after the Chimera of classical Greek mythology which had the head of lion, the body of a goat and the tail of a snake. A species has been described as a group of organisms which can mate and produce fertile offspring. A dog cannot mate with a cat; even if they did the offspring would not be fertile. A horse and the donkey are not the same species. Although they can mate, the offspring the mule, is not fertile.

Genetic engineering has enabled the crossing of the species barrier, in that DNA from one organism can now be introduced into another where such exchange would not be possible under natural conditions. With this technology engineered cells are now capable of producing metabolic products vastly different from those of the unaltered natural recipient.

Procedure for the Transfer of the Gene in Recombinant DNA Technology (Genetic Engineering):

- In broad items the following are the steps involved in in vitro recombination or genetic engineering. The bulk of the work done so far has been with E. coli as the recipient organism
- Dissecting a specific portion from the DNA of the donor organism.
- Attachment of the spliced DNA piece to a replicating piece of DNA (or vector), which can be from either a bacteriophage or a plasmid.
- Transfer of the vector along with the attached DNA (*i.e.*, the DNA chimera) into the host cell.
- Isolation (or recognition) of cells successfully receiving and maintaining the vector and its attached DNA.

Dissection of a Portion of the DNA of the Donor Organism

The donor DNA may come from a plant, an animal, a microorganisms or may even be synthesized in the laboratory. The dissection of DNA at specific sites is done by enzymes obtained from various bacteria and known as restriction endonucleases.

Nature and Types of Restriction Endonucleases

Restriction endonucleases are nucleic acid-splitting enzymes and are termed 'restriction' because they help a host cell destroy or restrict foreign DNA which enter the cell.

The host protects its DNA from its own restriction endonucleases by the introduction of methyl groups at recognition sites where the cleavage of the DNA occurs. The host DNA so protected is said to be 'modified. ' For every restriction enzyme there is a modification one hence the enzymes exist as restriction-modification complexes.

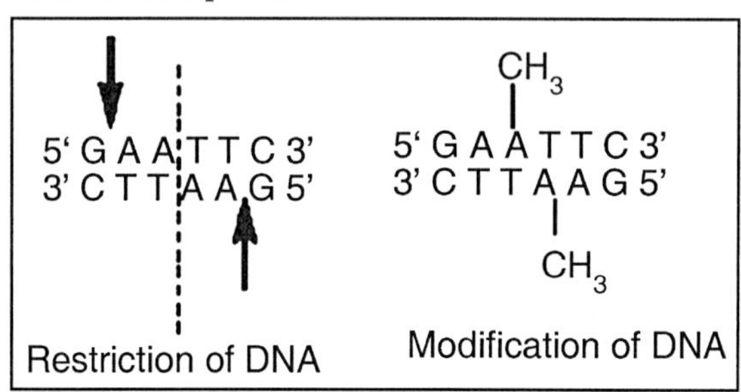

Their discovery was an important landmark in molecular biology. Daniel Nathans and Hamilton Smith received the 1978 Nobel Prize in Physiology and Medicine for their isolation of restriction endonucleases, which are able to cut DNA at specific sites. Conventionally restriction enzymes are denoted as the single stranded DNA; the position of the restriction is written/while the position of the modification is written as an asterisk.

Thus the representation for the above enzyme would 5'G/AA*TTC3'. There are four different types of restriction endonucleases: Types I, II, III and IV (Type IV is designated Type II S by some authors), but only Type II is used extensively in gene manipulations.

In Types I and III, one enzyme is involved for recognition of specific DNA sequences for cleavage and methylation, but the cutting positions are at variable distances from these sites (sometimes up to 1000 base pairs (bps)) away from these sites.

Type IV cuts only methylated DNA. As most molecular biology work is done with Type II endonucleases and only they will be discussed.

Type II endonucleases have the following advantages over the others. Firstly in Type II systems, restriction and modification are brought about by different enzymes and hence it is possible to cut DNA in the absence of modification (note that in Types I and III a single enzyme is involved); secondly, Type II enzymes are easier to use because they do not require enzyme cofactors.

Finally as will be seen below they recognize a defined symmetrical sequence and cut within this sequence. Type II restriction endonucleases recognize and cut DNA within particular sequences of 4 to 8 nucleotides in an axis of symmetry in such a way that the sequences of the top strand when read backwards are exactly like the bottom on the other side of the axis thus:

5'– ATG | CAT – 3'
3'– TAC | GTA – 5'

Axis of symmetry

Such sequences are referred to as palindromes. Type II restriction endonucleases were discovered in Haemophilus influenzae in 1970. About 3, 000 of theses enzymes have now been discovered and they cut in about 200 patterns; many of them are available commercially.

Nomenclature of Restriction Endonucleases

The nomenclature of restriction endonucleases is based on the proposals of Smith and Nathans and the currently adopted procedure is as follows:

- The species name of the host organisms is identified by the first letter of the genus name and the first two letters of the species name to form a three-letter abbreviation written in italics. For example, E. coli is Eco and Haemophilus inflenzae, Hin.
- Strain or type identification is supposed to be written as a subscript. Thus, E. coli strain K, EcoK. In practice it is all written in one line Ecok.
- Where a particular host has several different restriction and modification systems, these are identified by Roman numerals. Thus, those from H. influenzae strain Rd. would be Hind I, Hind II, Hind III, in the order of their discovery.
- Restriction enzymes have the general name endonuclease R and in addition carry the system name, thus endonuclease R. Hind I. Modification enzymes are named methylase M; thus the modification enzyme from H. influenzae Rd. is named methylase M. Hind I. Where the context makes it clear that restriction enzymes are being discussed, 'endonuclease R' is left out leaving Hind I.

Cutting DNA by Type II Endonucleases

Type II endonucleases recognize and break DNA within particular sequences of four, five, six, or seven nucleotides which have a symmetry along a central axis.

The same restriction endonuclease is used to cut the foreign DNA to be inserted into a vector, as well as the vector itself, in order to open it up. Restriction enzymes cut DNA between deoxyribose and phosphate groups, leaving a phosphate at 5' end and an OH group at the 3'.

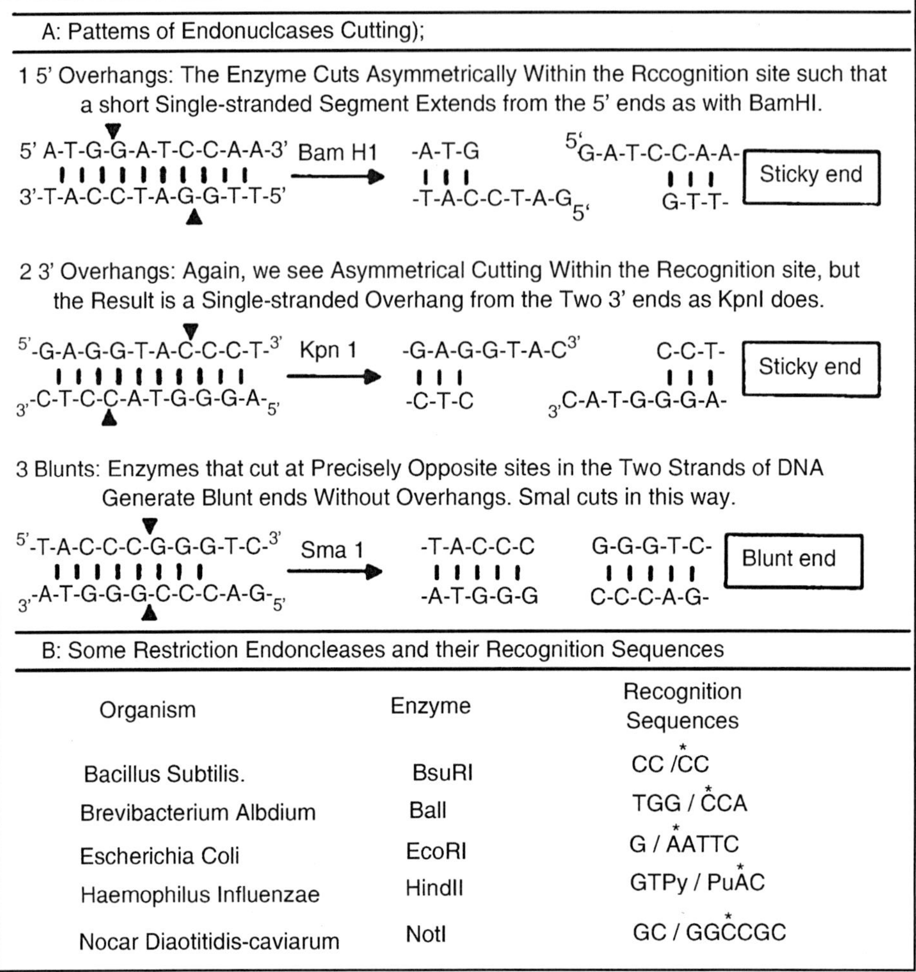

Organism	Enzyme	Recognition Sequences
Bacillus Subtilis.	BsuRI	CC /C*C
Brevibacterium Albdium	BalI	TGG / C*CA
Escherichia Coli	EcoRI	G / A*ATTC
Haemophilus Influenzae	HindII	GTPy / PuA*C
Nocar Diaotitidis-caviarum	NotI	GC / GGC*CGC

The Restriction enzymes used in genetic engineering cut within their recognition sites and generate one of three types of ends:

1. Single-stranded, "sticky" or cohesive ends as cut by Bam H1 (1, 5' overhangs).
2. Single-stranded, "sticky" or cohesive ends as cut by Kpn 1 (2, 3' overhangs).
3. Double-stranded, "blunt" ends as cut by Sma 1.(3, Blunts)

The single-stranded sticky or cohesive ends of DNA ends will join (anneal) with any DNA with sticky ends, having complimentary bases no matter the origin of the.

DNA, provided that both DNA samples have been cut with the same restriction enzyme.

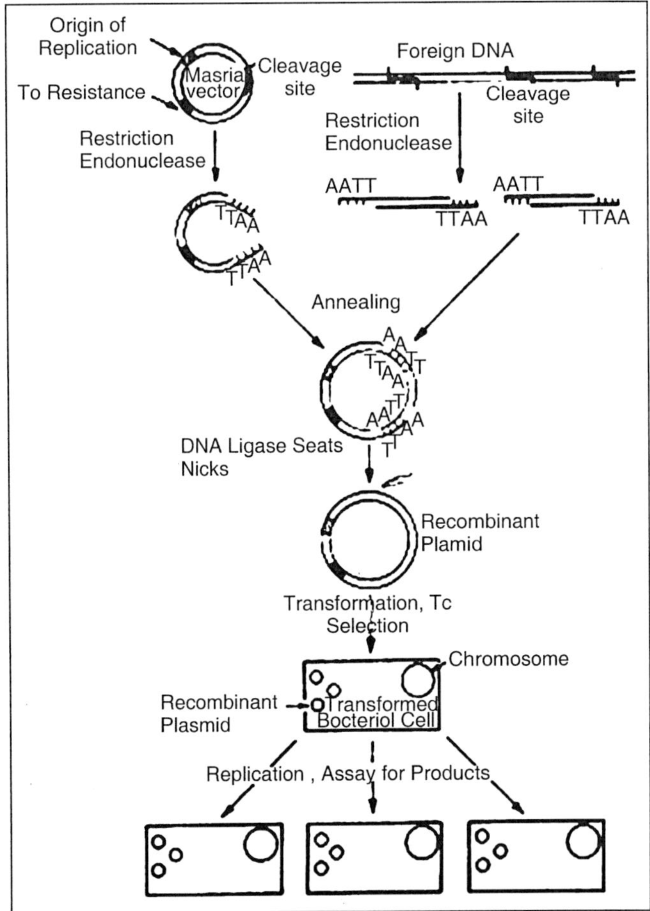

Fig. Generalized Diagram of Procedure for Genetic Engineering

The Attachment of the Spliced Piece of DNA to a Vector

- *Joining DNA molecules*: Three methods are used for the in vivo 'tying' of DNA molecules. The first method uses an enzyme DNA ligase to tie sticky ends produced by restriction endonucleases; the second is the use of another DNA ligase produced by E. coli infected by T4 bacteriophages to link blunt ended DNA fragments. The third method uses an enzyme terminal deoxynucleotidyl–transferase isolated from calf thymus to introduce single-stranded complimentary tails to two different DNA populations after which they anneal when mixed. Only the first, method, *i.e.*, the use of DNA ligase, will be discussed, because this has been used extensively.
- *The use of DNA ligase to join foreign DNA to the vector*: High concentrations of the DNA of the previously circular vector (usually a plasmid) and of the foreign DNA to be cloned onto the vector, are

mixed. Both DNA types have sticky ends having been treated with the same restriction endonuclease: in the case of the foreign DNA to cut it from its source and in the case of the vector, to open it up. Complimentary sticky ends from the foreign DNA and the vector anneal leaving however gaps created by the absence of a few base pairs in opposite strands.

- The enzyme DNA ligase can repair these gaps to create an intact duplex. DNA ligase is produced by E. coli and phage T4. The ligase from T4 can, however, join blunt-ended DNA whereas that from E. coli cannot. The vector-foreign DNA chimaera is then introduced into the bacterial cell by transformation.
- To prevent recircularization of the linearized vector, it may be treated with alkaline phosphotase. When it is so treated circularization can only occur when a foreign DNA is introduced. A gap is left at each joint. These gaps are closed after transformation by the hosts' repair system.
- *Vectors used in recombinant DNA work*: Two broad groups of cloning vehicles have been used, namely plasmids and lamda phages. Both have replication systems that are independent of that of the host cell.

Plasmids

Plasmids are circular DNA molecules with molecular weights ranging from a few million to a few hundred million Daltons. Plasmids appear to be associated with virtually all known bacterial genera. They replicate within the cell.

Some of the larger plasmids, known as conjugative plasmids, carry a set of genes which promote their own transfer in a sexual process known as conjugation which has already been discussed. Smaller plasmids are usually non-conjugative but their transfer can usually be promoted by the presence of a conjugative plasmid in the same cell.

Besides genes for sexual transfer, plasmids usually carry genes for antibiotic or heavy metal resistance. They often also carry genes for the production of toxins, bacteriocins, antibiotics, and unusual metabolites. In some cases they may carry genes for unusual capabilities such as the breakdown of complex organic compounds. Plasmids are, however, not essential for the cell's survival. Two important features of plasmids to be used in genetic experiments may be compared by examining two plasmids. Plasmid psC101 has only two to five copies per cell and replicates with its host DNA. It is said to be under 'stringent' control. However, another plasmid pCol E 1 is found in about 25-30 copies per cell. It has a 'relaxed control' independent of the host and replicates without reference to the host DNA.

When the host cell is starved of amino-acids or its protein synthesis is inhibited in some other manner, such as with the use of chloramphenicol, the Col E 1 plasmid continues to replicate for several hours until there are 1, 000 to 3, 000 copies per cell. Due to this high level of gene dosage (also referred to as gene amplification), products synthesized because of the presence of these plasmids are produced in extremely high amounts, a property of immense importance in biotechnology and industrial microbiology.

Generally conjugative plasmids are large, exhibit stringent control of DNA replication, and are present in low copy numbers; on the other hand, non-conjugative plasmids are small, show relaxed DNA replication, and are present in high numbers. Many other plasmid vectors exist, some constructed in the laboratory.

Table. Some Commonly used Plasmid Cloning Vehicles

Plasmid	Molecular Weight (x 10^{-6})	Marker*	Single Restriction Sites
pSC101	5.8	Tc	BamHI, EcoRI, HindIII, Hpal, SalI, Smal
Col E1	4.2	Colimm	EcoRI
pMB9	3.6	Tc^r, Colimm	BamHI, EcoRI, HindIII, Hpal, SalI Smal
pBR313	5.8	Tc^r, Ap^r Colimm	BamHI, EcorI, HindIII, Hpal, SalI, Smal
pBR322	2.6		BamHI, EcoRI, HindIII, pstI, SalI

Note:
**Tcr*: tetracycline resistance. Apr: ampicillin resistance
colimm: colicin immunity

Ideal Properties in a Plasmid used as a Vector

A plasmid to be used in genetic engineering should ideally have the following properties:

- The plasmid should be as small as possible so the unwanted genes are not transmitted, as well as to facilitate handling;
- It should have an origin of replication, the site where DNA replication initiates;
- It should have a relaxed mode of replication;
- It should have sites for several restriction enzymes;
- It should carry, preferably, two marker genes. Marker genes are those which express characteristics by which the plasmid can be identified. Such characteristics include resistance to one or more antibiotics. A marker of great importance is the ability to satisfy auxotrophy, *i.e.*, the ability to produce an amino acid or other nutritional component which the host's chromosome is incapable of producing.
- The nucleotide sequence of the plasmid should be known;
- For safety reasons the plasmid should not be able to replicate at mammalian body temperatures so that should it enter the human body and be able to produce deleterious substances, it should fail to replicate;
- For safety reasons also, it should not be highly transmissible by conjugation if it controls the production of any material harmful to the mammalian body;

- The plasmid as a cloning vehicle should have a site for inducing transcription across the inserted fragment. The plasmid-initiated transcription should be controlled by the host (by induction or repression). Uncontrolled transcription could be harmful to the host.

Some commonly used plasmid cloning vehicles. They carry various markers based on tetracycline or ampicillin resistance or immunity against colicin attack.

The marker may be carried either on the plasmid or on the inserted DNA. If neither of them carries a marker then DNA carrying a marker can be grafted on to either the vector or the insert.

Plasmids Currently in use for Cloning

In the early years of genetic engineering, naturally occurring plasmids such as Col E1 and pSC 101 were used as cloning vectors. They were small and had single sites for the common endonucleases. However they lacked markers which would help select transformed organisms. New plasmids were therefore developed.

The best and most commonly used is pBR322 developed by Francisco Bolivar. (In naming plasmids p is used to show it is a plasmid; p is followed by the initials of the worker who isolated or developed the plasmid; numbers are used to denote the particular strain).

Plasmid pBR322 has all the properties expected in a plasmid vector: low molecular weight, two markers, (resistance to ampicillin, ApR and tetracycline, TcR) an origin of replication, and several single-cut replication sites. Modifications of the original pBR322 have been made to suit special purposes, and consequently many variants exist in the pBR322 family.

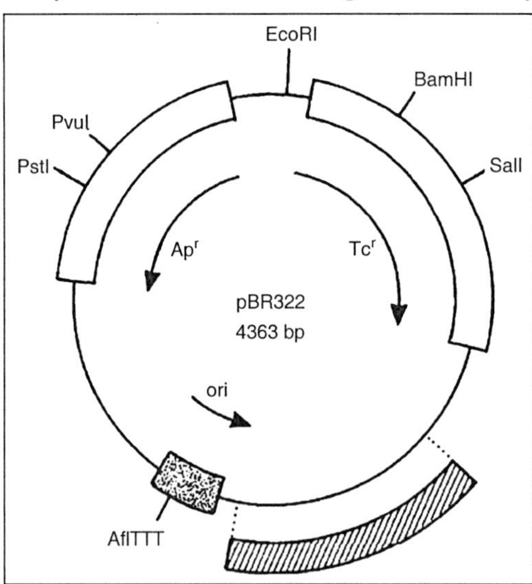

Fig. Genetic Map of Plasmid pBR322 Showing Unique Recognition Endonuclease Sites and Genes for Tetracycline and Ampicillin Resistance

A widely used variant of pBR322 is pAT153, which some consider a better vector than its parent because it is present in more copies per cell than pBR322 Another series of popular vectors is the pUC family of vectors.

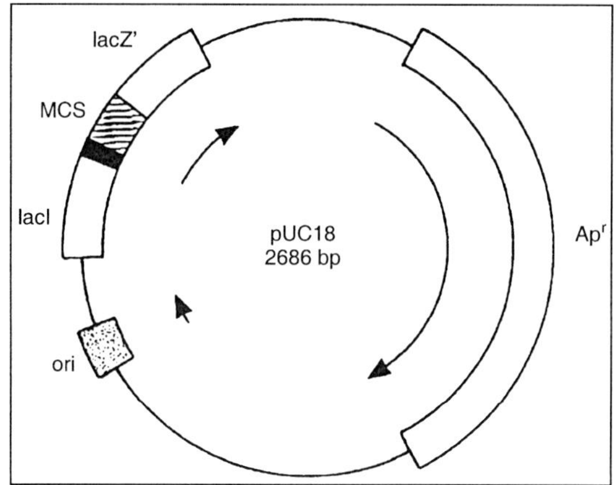

Fig. Genetic Map of pUC18

It has several unique restriction sites in a short stretch of DNA, which is an advantage in some kinds of work.

Phages

Two types of phages have been developed for cloning, _ lamda, and M13. Most of the phages used for cloning are derivatives of the lamda phage of E. coli because so much is already known about this phage. Derivatives are used because the wild-type phage is not suitable as a vector as it has several targets of sites for most of the most commonly used endonucleases. The chromosome of phage must be folded and encapsulated into the head of the virus in order to provide a mature virion.

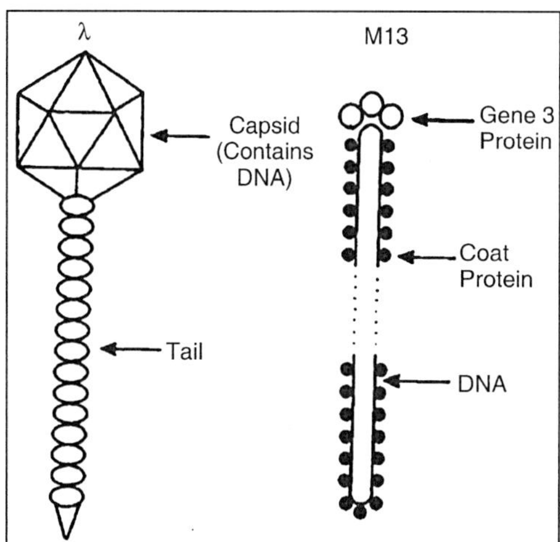

Fig. Structure of l and M13 Bacteriophages

The amount of DNA that can enter the head is limited, and hence the available DNA in a phage is also limited. Therefore unwanted phage DNA must be removed as well as all but one of restriction targets for the chosen enzyme. The DNA of phage lambda when it is isolated from the phage particle is linear and double-stranded.

At each end of the chain are single-stranded portions which are complimentary to each other, much like the 'sticky ends' produced from DNA cutting by restriction endonucleases.

These lamba DNA pieces are able to circularize and replicate independently within the host.

The middle portion of the linear double-stranded phage DNA is non-essential for phage growth and it is here that the foreign DNA is introduced. The more distal positions carry genes which code for essential components such as the head, tail of the bacteriophage and the host lysis

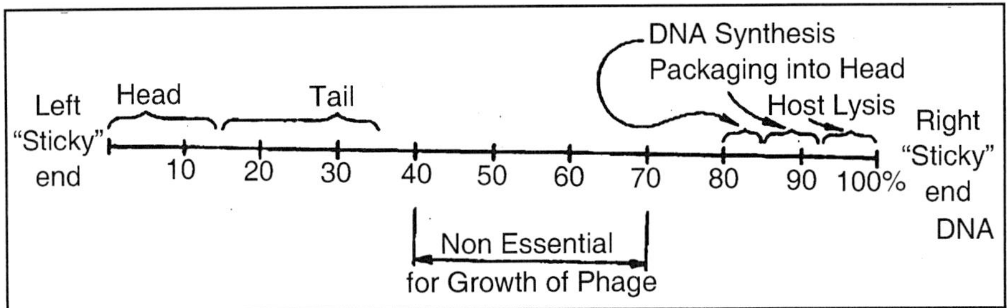

- *Transfection*: The linear chimera can be introduced by transformation. (When virus DNA is transformed the process is known as transfection.) However, much of introduced chimeras are restricted in comparison to when pure phage DNA is transfected.
- *Packaging the chimeras into virus heads*: The recombinant DNA or chimera may be packaged into a virus head and a tail attached by in vitro means. The procedure for this packaging is outside the scope of this book but may be found elsewhere. Once packaged, the synthetic virus can then inject its DNA into the host in the usual way.

Cosmids

Cosmids are plasmids constructed from phage DNA by circularization at the 'sticky, ' single stranded ends or cos sites. Foreign DNA is attached to the cosmid which is then packaged into a phage. When the cosmid is injected it circularizes like other virus DNA but it does not behave as a phage, rather it replicates like a plasmid. Drug resistance markers carried on it help to identify it.

A number of commercially available vectors are based on phages and some are shown in Table.

Table. Phage-based Vectors

Vector	Features	Applications
pBR322	Ap^rTc^r Single cloning sites	General cloning and sub-cloning in E. coli
pAT 153	Ap^rTc^r Single cloning sites	General cloning and sub-cloning in E. coli
pGEM™-32	Ap^r MCS SP6/T7 promoters lacZα-peptide	General cloning and in vitro transcription in E. coli and mammalian cells
pCI™	Ap^r MCS T7 promoter CMV enhance/promoter	Expression of genes in mammalian cells
pCMV-Script™	Neo^r Large MCS CMV enhancer/promoter	Expression of genes in mammalian cells Cloning of PCR products

Transfer of Vector along with the Attached DNA into the Bost all

The vector once spliced with the endonuclease cannot reform into a circular structure unless a suitable fragment of the foreign DNA with a complimentary 'sticky end' fits in. Foreign DNA digests produced by physical inactivation may also be used.

If a large enough amount of foreign DNA digest is used the probability is that a piece with the appropriate complementary end will fit in. The new hybrid DNA is introduced into the host cell by transformation. Transformation is facilitated by treating the host cells in calcium salts, after washing them in magnesium salts.

Recognizing the Transformed Cell

The introduction of the new property into the host may be detected by growing the cells in a medium containing antibiotics whose resistance is specified in the introduced foreign DNA. Growth should occur if the resistance gene was transferred. If genes for the synthesis of some products were introduced via the chimera, the transformed bacteria should grow on the selective medium. The products are then examined for the synthesis of the new compound. For the introduced gene to lead to protein synthesis a suitable promoter must be present; it may be introduced from other organisms if an appropriate one is not indigenous.

Gene Transfer into Organisms other than E. coli, including Plants and Animals

The methods discussed are those developed primarily for E. coli. The discussion will look at the introduction of DNA into bacteria other than E. coli as well into other organisms, including plants and animals. Some of the methods to be discussed are also used on E. coli.

Delivery into Bacteria other than E. coli

- *Electroporation*: In the process of electroporation, cells into which DNA is to be introduced (*i.e.*, cells to be transfected) are exposed to high-voltage electric pulses. This creates temporary holes in the cell membrane through which DNA can pass. Electroporation can be used for transfecting cells of Bacilli spp and actinomycetes, especially when protoplasts are produced from the cells. Electroporation is the short form for electric field-mediated membrane permeabilization. It is still used for E. coli, especially when chimera longer than about 100 kilobases (100 kb) are to be used. In general electroporation can be used for transfecting bacteria and archae, after the appropriate electric voltage and other parameters have been worked out. It is also used for transfecting plant cells.
- *Conjugation*: In some bacteria where other means of introducing DNA appear difficult or have failed, the natural means of transferring DNA by plasmid mediated conjugation has been exploited. A conjugative plasmid which is carrying the insert and which has the genes for its own transfer is used. However where this is not possible, a conjugative plasmid with its own transfer gene may first be introduced, followed by the non-conjugative (*i.e.*, does not promote the transfer of DNA through pili) plasmid carrying the DNA insert. This has been used in some strains of Pseudomonas.
- *Use of Liposomes*: When the DNA to be introduced is first entrapped in phospholipid droplets known as liposomes, it enhances the entry of the DNA into protoplasts of Gram-positive Bacillus and actinomycetes. Liposomes have also been used for delivering DNA into animal cells. Liposomes are microscopic, fluid-filled vesicles whose walls are made of layers of phospholipids identical to the phospholipids that make up cell membranes. The outer layer of the vesicle is hydrophobic, while the inner layer is hydrophilic; this enables the liposome to carry water soluble materials within it. They can be designed so that they have cationic, anionic or neutral charges at the hydrophobic end depending on the purpose for which they are meant. They are used for introducing DNA into animal, plant or bacterial cells. When used for introducing DNA into plant cells, such cells must have their cell walls removed, yielding protoplasts; with bacteria, the cell walls must also be removed to yield sphaeroplasts. Liposomes have been used experimentally to carry normal genes into a cell to replace defective, disease-causing genes in gene therapy. Liposomes are used to deliver certain vaccines, enzymes, or drugs

(*e.g.*, insulin and some cancer drugs) to the animal body. Liposomes are sometimes used in cosmetics because of their moisturizing qualities.

Delivery of DNA into Plant Cells

Plants are peculiar in that most single plant cells can be caused to develop into the entire plants. Successfully transfecting (*i.e.*, introducing foreign DNA into) a plant cell will result in having the foreign DNA as part of the genetic apparatus of the transfected plant. The introduction of foreign DNA into plants is done for the improvement of the agricultural, ornamental, nutritional, or horticultural value of the plant. It is also done to convert the plants into 'living fermentors' which with the appropriate genes can manufacture cheaply, some industrially important materials, which it may not even be possible to produce by chemical means. Several methods are available for the delivery of

DNA into plants:

- *Plant Transfection with Ti plasmid of Agrobacterium tumefaciens*: The Gram-negative soil bacterium, Agrobacterium tumefaciens, is the causative agent of crown gall, a disease which produces tumors in plants (mainly dicots) on entering through a wounded plant cell. The pathogenic properties of the bacterium are due to the Ti (tumor inducing) plasmid which it carries. Part of the Ti plasmid, known as the T (transfer) DNA, is transferred to the plant cell and is integrated into the genome of the host under the direction of the virulence gene.

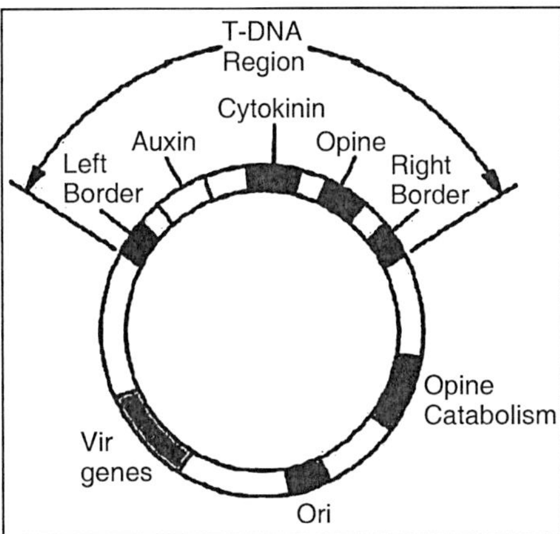

Fig. Map of Agrobacterium Tumefaciens Ti Plasmid

It is within the TDNA that foreign DNA can be introduced. A section of the TDNA codes for the production of auxins and cytokins which lead to the formation of galls or tumors. Another section codes for the production of conjugates of amino acids and sugars known as opines and which are metabolized by Agrobacterium tumefaciens residing in the tumor. The oncogenic (tumor-causing) and the opine-producing portions of the TDNA of wild-type Agrobacterium tumefaciens are removed, when it is to be used for cloning. Furthermore, marker genes (*e.g.*, for kamycin resistance) are introduced into the plasmids so that transformed plants can be identified. Because the marker genes are of bacterical origin an origin of replication from E. coli is also introduced. The TDNA is defined by the left and right borders. The sequences of the right border are essential for the TDNA transfer and integration into the host plant. The transfected plant cells therefore result in normal plants. Ti plasmids in which the oncogenic section has been removed is said to be 'disarmed'. Such disarmed plasmids lack the sequences necessary to produce the phytohormones which give rise to diseased conditions, gall or tumor. Other properties such as the transfer of DNA are still active and the regeneration of healthy plants can still occur. The Agrobacterium tumefaciens Ti plasmid has been successfully and widely used in cloning in plants. However it has been more successful in dicots than in monocots.

- *Use of Viruses*: The cauliflower mosaic virus (caMV) has been used as a powerful vector for introducing DNA into plants. It is a double-stranded DNA virus with 8025 base pairs. Certain portions of the virus are dispensable and foreign genes can be replace them. However, it has a limited host range; furthermore foreign sequences are often unstable in the caMV genome.
- *Electroporation*: Electroporation is widely used for transfecting plant cells. When plants are to be transfected, protoplasts or whole plant cells placed in contact with exogenous DNA in foil-lined cuvettes and exposed to high electrical current. The cells become permeable and take up exogenous DNA, some of which integrate with the plant genome. It has been successfully used in a wide variety of species using equipment which is relatively inexpensive.
- *Biollistic or Microprojectile methods*: This is one of the commonest methods used for transfecting plants. In this method a socalled gene gun or particle gun is used to shoot tiny pellets of tungsten or gold coated with the foreign DNA in question into the leaves or stem of the plant to be transformed. It is a widely used and highly successful method of transfecting plants using plant protoplasts, plant cell suspensions, callus cultures, even chloroplasts and mitochondria, and indeed any form of plant preparation capable of regeneration in dicots, monocots, and conifers. Success occurs more with linear DNA than with circular; furthermore very large DNA inserts tend to be broken during the projection. When inside the cell, some of the introduced DNA get integrated with the plant DNA.

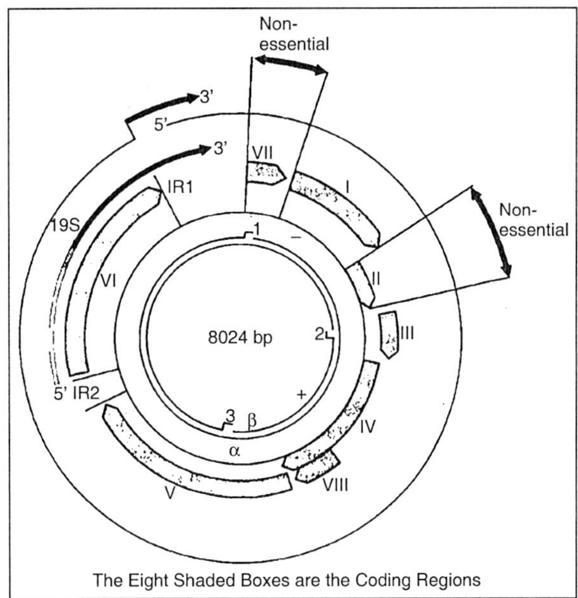

Fig. Genetic Map of the Cauliflower Mosaic Virus

- *Microinjection*: This method involves immobilizing the cells and injecting DNA into protoplasts, walled cells, or embryos. It is done with a fine needle under the microscope. The technique needs a lot of skill. Some authors do not think it has much future because only one cell can be injected at a time.

Delivery of DNA into Animal cells

Genetic engineering in plants differs in at least two respects from that in animals. Firstly while plant cells are mostly totipotent (*i.e.*, most plant cells are able to give rise to a new plant), animal cells cannot give rise to whole animals once differentiated into specialized cells. In animals the cells that become reproductive cells separate early from those that are ordinary body (somatic) cells.

Somatic cells do not give rise to new animals To create transgenic animals the foreign DNA must be introduced into cells while they are still totipotent and differentiation has not occurred. Generally this involves introducing the DNA into stem cells (yet undifferentiated cells), an egg, the fertilized egg, (oocyte or zygote) or early embryo.

Some of the methods discussed for introducing foreign DNA into bacteria and plants are also applicable to animal cells: electroporation, biollistic methods and microinjection have all been successfully used in animals. In addition the liposome (phospholipid) delivery seen in bacteria is also used in animal cells. Genes are introduced into animal cells as well as in vivo by transduction via viruses in gene therapy. Four groups of viral vectors are used for gene therapy in humans: adenoviruses, baculoviruses, herpesvirus vectors, and retroviruses. Changing the genetic make-up of animals, in large domesticated mammals such as cows, pigs and sheep, allows a number of commercial applications.

These applications include the production of animals which express large quantities of exogenous proteins in an easily harvested form (*e.g.*, expression into the milk), the production of animals which are resistant to infection by specific microorganisms and the production of animals having enhanced growth rates or reproductive performance.

Most of the work on transgenic animals has been done with mice on account of their small size and low cost of housing in comparison to that for larger vertebrates, their short generation time, and their fairly well defined genetics.

Foreign DNA is introduced in mice in one of the following ways: DNA microinjection, embryonic stem cell-mediated gene transfer and retrovirus-mediated gene transfer, sperm-mediated transfer, transfer into unfertilized ova.

- *DNA microinjection*: This method involves the direct microinjection of a chosen gene construct (a single gene or a combination of genes) from another member of the same species or from a different species, into the pronucleus of a fertilized ovum. The introduced DNA may lead to the over- or under-expression of certain genes or to the expression of genes entirely new to the animal species. The insertion of DNA is, however, a random process, and there is a high probability that the introduced gene will not insert itself into a site on the host DNA that will permit its expression. The manipulated fertilized ovum is transferred into the oviduct of a recipient female, or foster mother that has been induced to act as a recipient by mating with a vasectomized male. Such males cannot inject sperms into the female because the tubes carrying the sperms, the vas deferens, have been cut. The major advantage of this method is its applicability to a wide variety of species.
- *Embryonic stem cell-mediated gene transfer*: This method involves prior insertion of the DNA sequence by homologous recombination into an in vitro culture of embryonic stem (ES) cells. Stem cells are undifferentiated cells that have the potential to differentiate into any type of cell (somatic and germ cells) and therefore to give rise to a complete organism. These cells are then incorporated into an embryo at the blastocyst stage of development. The result is a chimeric animal. ES cellmediated gene transfer is the method of choice for gene inactivation, the so-called knock-out method. This technique is of particular importance for the study of the genetic control of developmental processes. This technique works particularly well in mice. It has the advantage of allowing precise targeting of defined mutations.

- *Retrovirus-mediated gene transfer*: To increase the probability of expression, gene transfer is mediated by means of a carrier or vector, generally a virus or a plasmid. Retroviruses are commonly used as vectors to transfer genetic material into the cell, taking advantage of their ability to infect host cells in this way. Offspring derived from this method are chimeric, *i.e.*, not all cells carry the retrovirus. Transmission of the transgene is possible only if the retrovirus integrates into some of the germ cells.
- *Sperm-mediated Gene Transfer*: Sperms may be coated with the target DNA or attached to the sperm through a linker protein, and introduced through surgical oviduct insemination. It has been successfully used in pigs. With the above techniques the success rate in terms of live birth of animals containing the transgene is extremely low. If there is birth, the result is a first generation (F1) of animals that need to be tested for the expression of the transgene. The F1 generation may result in chimeras. When the transgene has integrated into the germ cells, the germ line chimeras are then inbred for 10 to 20 generations until homozygous transgenic animals are obtained and the transgene is present in every cell. At this stage embryos carrying the transgene can be frozen and stored for subsequent implantation.

Application of Genetic Engineering in Industrial Microbiology and Biotechnology in General

The unparalleled ability of DNA to replicate and reproduce itself is truly remarkable. What this means is that, put crudely, DNA of a given sequence coding for the production of a polypeptide or protein in organism A will lead to the production of the same polypeptide or protein if the same sequence is put into organism B.

This is the basic assumption underlying the numerous advances in our manipulation of the biotic world for the benefit of humans. This section looks only at some of the numerous positive changes recombinant DNA technology has contributed to spreading a better quality of life to millions of people around the world through improvements in agriculture, health care delivery and industrial productivity.

- Production of Industrial Enzymes: Genetically engineered bulk enzymes are used mostly in the food industry (baking, starch manufacture, fruit juices), the animal feed industry, in textile manufacture, and in detergents. A leading manufacturer of these enzymes among world manufacturer is Novo Enzymes of Denmark.

  The Advantages of using Engineered Enzymes are as follows:
  - Such enzymes have a higher specificity and purity;
  - It is possible to obtain enzymes which would otherwise not be

available due to economical, occupational health or environmental reasons;

- On account of the higher production efficiency there is an additional environmental benefit through reducing energy consumption and waste from the production plants;
- For enzymes used in the food industry particular benefits are for example a better use of raw materials (juice industry), better shelf life of the final food and thereby less wastage of food (baking industry) and a reduced use of chemicals in the production process (starch industry);
- For enzymes used in the animal feed industry particular benefits include a significant reduction in the amount of phosphorus released to the environment from farming.
- Two enzymes will be discussed briefly: chymosin (rennets) and bovine somatotropin (BST). Chymosin is also known as rennets or chymase and is used in the manufacture of cheese. It used to be produced from rennets of farm animals, namely calves. Later it was produced from fungi, Rhizomucor spp. Over 90 per cent of the chymosin used today is produced by E. coli, and the fungi, Kluyveromyces lactis and Aspergillus niger. Genetically engineered chymosin is preferred by manufacturers because while it behaves in exactly the same way as calf chymosin, it is purer than calf chymosin and is more predictable. Furthermore, it is preferred by vegetarians and some religious organizations.

Table. Some Genetically Engineered Industrial Enzymes

Type of enzyme	Main application
Alpha-amylase/Bacillolysin/Xylanase baking industry	Brewing industry, starch industry,
Amyloglucosidase	Alcohol industry, fruit processing
Cellulase	Detergent industry; textiles
Decarboxylase	Brewing industry
Glucoamylase	Alcohol industry, starch industry
Glucose oxidase	Baking industry
Lipase	Oils and fats industry; baking industry; dairy industry; leather industry
Lipase	Pasta/noodles
Maltogenic amylase	Starch industry, baking industry
Pectate lyase	Textile industry, fruit processing
Pectinesterase	Fruit processing
Phytase	Animal feed industry
Protease	Meat industry; detergent industry
Pullulanase/Amyloglucosidase	Starch industry, fruit processing

- Bovine Somatotropin (BST) is a growth hormone produced by the pituitary glands of cattle and it helps adult cows produce milk. It is produced by genetic engineering in E. coli using a plasmid vector. Supplementing dairy cows with bovine somatotropin safely enhances milk production and serves as an important tool to help dairy producers improve the efficiency of their operations. The use of supplemental BST allows dairy farmers to produce more milk with fewer cows, thereby providing them with additional economic security. It is marketed by Monsanto as Posilac.

- Enhancing the activities of Industrial Enzymes: Through protein engineering it has been possible to enhance the properties of proteins to make them more stable to denaturation, more active in their biocatalytic ability and even to design new properties in existing enzymes. The properties of proteins are due to their conformation which is a result of their amino acid sequence. Certain amino acids in a protein play important parts in determining the stability of the protein to high temperatures, specificity and stability to acidity. In protein engineering changes are caused to occur in the protein by changes in the nucleotide sequence; a change of even a single nucleotide could lead to a drastic change in a protein. Many industrial processes are carried out at elevated temperatures, which can unfold the proteins and cause them to denature. The addition of disulphide bonds helps to stabilize them. Disulphide bonds are usually added by engineering cysteine in positions where it is desired to have the disulphide bonds. The addition of disulphide bonds not only increases stability towards elevated temperatures, but in some instances also increases stability towards organic solvents and extremes of pH. An example of the increase of stability to elevated temperatures due to the addition of disulphide bonds is seen in xylanase. Xylanase is produced from Bacillus circulans. During paper manufacture, wood pulp is treated with chemicals to remove hemicelluloses. This treatment however leads to the release of undesirable toxic effluents. It is possible to use xylanase to breakdown the hemicellulose. However, at the time when bleaching is done, the pulp is highly acidic as a result of the acid used to digest the wood chips to produce wood pulp. The acid is neutralized with alkali, but the temperature is still high and would denature native xylanase. In silico (*i.e.* computer) modeling showed the sites where disulphide bonds can be added without affecting the enzyme's activity. The introduction of the disulphide bonds did increase the thermostability of the enzyme, making it possible to keep 85 per cent of its activity after 2 hours at 60°C whereas the native enzyme lost its activity after about 30 minutes at the same temperature. Another way in which enzyme activity can be enhanced by protein engineering is to actually increase the activity of the enzyme. This can be done only with an enzyme whose conformation, including the active sites, is thoroughly understood. Using in silico modeling, it is possible to predict the effect of changing amino acids at the active site of an enzyme. This has been done with the enzyme tRNA synthase from Bacillus stearothermophilus. Various other properties have been engineered into proteins including a modification of the metal co-factor and even a change in the specificity of enzymes.

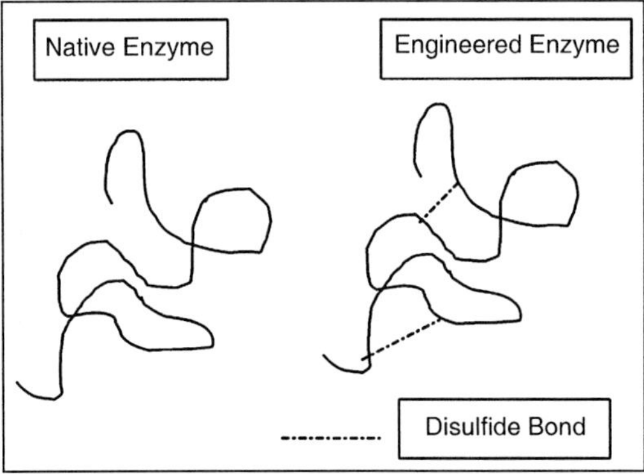

Fig. Stabilizing Enzymes through the Introduction of Disulfide Bonds

- Engineered Products or Activities Used for the Enhancement of Human Health Engineered health care products and activities can be divided into:
  - Those used to replace or supplement proteins produced by the human body in insufficient quantities or not produced at all;
  - Those involving the replacement of a defective gene;
  - Those that are used to treat disease,
  - Those that are used for prophylaxis or prevention of disease, *i.e.*, vaccines, or
  - Those that are used for the diagnosis of disease. Only insulin and edible vaccines will be discussed.
- *Insulin*: Insulin is a hormone produced by the pancreas; hormones are small proteins. Insulin is used to treat diabetes of which there are there three types, only two of which are relevant to this discussion.
- Type 1 Diabetes (previously known as insulin-dependent diabetes) is an auto-immune disease where the body's immune system destroys the insulin-producing beta cells in the pancreas. This type of diabetes,

also known as juvenile-onset diabetes, accounts for 10- 15 per cent of all people with the disease. It can appear at any age, although common under 40, and is triggered by environmental factors such as viruses, diet or chemicals in people genetically predisposed. To live, people with type 1 diabetes must inject themselves with insulin several times a day and follow a careful diet and exercise plan.

Table. Some Genetically Engineered Health Related Products

Product	Application
Hormones	
Insulin	Treatment of diabetes
Human growth hormone (somatotropin)	Treatment of dwarfism
Follicle stimulating hormone	Treatment of some disorders of the reproductive system
Immune System Participants	
Tumor necrosis factor	Anti-tumor agent
Interleukin 2	Treatment of some cancers
Lysozyme	Anti-inflammatory agent
A-Interferon	Antiviral
Blood Components	
Erythropoeitin	Treatment of anemia and cancers
Tissue plasminogen activate	Dissolves blood clots
Factor VIII	Treating hemophilia
Enzymes	
Human DNase I	Treatment of cystic fibrosis

- Type 2 Diabetes (previously known as non-insulin dependent diabetes) is the most common form of diabetes, affecting 85-90 per cent of all people with the disease. This type of diabetes, also known as late-onset diabetes, is characterized by relative insulin deficiency. The disease is strongly genetic in origin but lifestyle factors such as excess weight, inactivity, high blood pressure and poor diet are major risk factors for its development. Symptoms may not show for many years and, by the time they appear, significant problems may have developed. People with type 2 diabetes are twice as likely to suffer cardiovascular disease. Type 2 diabetes may be treated by dietary changes, exercise and/or medications. Insulin injections may later be required.
- The Third type affects pregnant women, is less common, and will not be discussed.
- Genetically engineered insulin was the first major product of biotechnology. As insulin from some animals is similar to human insulin, beginning from the 1920s, insulin isolated from the pancreas of farm animals, mainly pigs and cows, was used to treat diabetes.

There were several problems with this product. First it takes several months for animals to mature and be ready to be slaughtered for their pancreas. This made animalbased insulin expensive since it was difficult to meet the demand. Furthermore such animal insulin caused immune reactions in some patients and a few became intolerant or resistant to animal insulin. For a more effective solution the then new technology of recombinant DNA was resorted to. In 1978, in the laboratory of Herbert Boyer at the University of California at San Francisco, a synthetic version of the human insulin gene was constructed and inserted E. coli. In 1982 Eli Lilly Corporation was granted approval for its genetically engineered insulin. Insulin is a small protein, and today's insulin is produced with a synthesized gene, which is expressed in a yeast.

- *Insulin consists of two amino acid chains*: the A peptide chain which is acidic and with 21 amino acids and the B peptide chain which is basic and has 30 amino acids. When synthesized the A and B chains are further linked by a 30 amino acid C peptide chain to produce a structure known as pro-insulin. Pro-insulin is cleaved enzymatically to yield insulin.

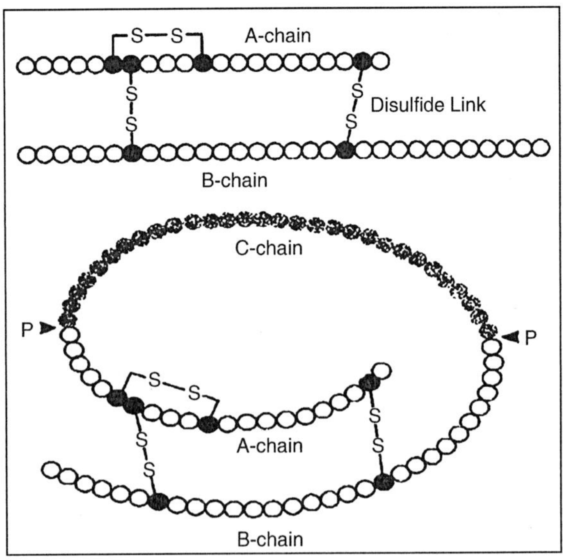

Fig. Structure and Synthesis of Insulin

Note:

Top: Structure of insulin. The A chain has 21 amino acids (represented by circles) and the B chain has 30 amino acids. The A and B chains are linked by disulfide bond between cysteine residues (filled circles) Bottom: Synthesis of Insulin: Proinsulin, the Precursor of Insulin

When synthesized proinsulin consists of an 81-amino acid polypeptide. The C chain is then cleaved off by a protease (P) to yield insulin.

- *Edible vaccines*: An innovative new approach to vaccine production is the surface expression of the antigen of a bacterium in a plant. Most current immunization is done by injection (parenteral delivery) and rarely results in specific protective immune responses at the mucosal surfaces of the respiratory, gastrointestinal and genito-urinary tracts. Mucosal immune responses represent a first line of defence against most pathogens. In contrast, mucosally targeted vaccines achieve stimulation of both the systemic as well as the mucosal immune networks. In addition, mucosal vaccines delivered orally increase safety and compliance by eliminating the need for needles. In addition many vaccines depend on the need of maintenance of a 'cold chain' (refrigeration) for delivery. Many developing countries, lack the resources to maintain the chain giving rise to many cases of vaccine failure, along with the fact they lack the resources and technology for fermentation industries. Both of these factors create constraints in vaccine use in the developing world, where these vaccines are needed the most. Combining a cost-effective production system with a safe and efficacious delivery system, plant edible vaccines, provide a compelling new opportunity.
  — Plant-based oral vaccines are cheap, safe and efficient. From the point of the little child receiving the vaccine, a smile might be elicited when a 'banana' vaccine is eaten rather the sharp cry of the pain of a needle! Vaccines have been produced in several plants, including a vaccine against dental carries caused by Streptococcus mutans, which was produced in the tobacco. Two other interesting examples are the expression of the rabies external antigen in tomatoes and the hepatitis virus antigen in lettuce.

Genetically Engineered Plants

Plants have been engineered for the introduction of many new desirable properties. ollectively these attainments represent a major triumph of biotechnology, enabling us to achieve in a few years what would take traditional plant breeding decades to attain, if at all.

Some genetic engineering achievements would be impossible with traditional plant breeding methods since in the latter, the introduction of new genetic properties occurs only through the exchange of sexual materials (in the pollen grains) of the same species.

In genetic engineering the natural species barrier is not recognized since the DNA sequence introduced into a plant can come from another plant of a different species or even from a non-plant source such as a bacterium, and indeed may even be synthesized. The introduction of some genetically engineered foods has met with public resistance, although many have been shown to be safe. What is required is continued public education about their safety before their introduction, and constant sensitivity to public opinion thereafter.

Table. Vaccines Produced in Plants

Vaccine Against	Plant
Cholera	Potato
Foot and mouth disease	Arabidopsis
Herpes virus	Tobacco
Norwark (diarrhea) virus	Potato
Rabies	Tobacco
Rabies	Tobacco

The ensuing discussion will be under two headings:

- Improving field, production or gronomic traits
- Modification of consumer products.

Improving Field or Production Characteristics

Numerous improvements have been made in agricultural crops by introducing into them enes coding for the desired properties, but only plant engineering for herbicide esistance, resistance to viral diseases, resistance to insect pests, and resistance to salt tress will be discussed.

Engineering Plants for Herbicide Resistance

An estimated US $10 billion is spent annually on weed killers. In spite of this about 10 per cent f world crop production is lost to weeds. Herbicides (weed killers) target processes that re essential and unique to plants. These processes are however important to plants and eeds alike, and getting methods that are selective for either is difficult. One method that s used is to engineer crops so they become resistant to the weed killer.

Plants can become esistant to herbicides in one of the four following ways:

1. Overproduction of the herbicide sensitive target, so that some is still left for the roper cell function despite presence of the herbicide in the cell;
2. Reduction of the ability of the herbicide-sensitive target protein to bind to the erbicide;
3. Engineering into plants the ability to metabolically inhibit the herbicide;
4. Inhibition of the uptake of the herbicide.

It is important to have some idea of the modes of action of herbicides so as to nderstand how the genetic engineering is to proceed.

The most common modes of action re given below and examples of herbicides in each group are given in parenthesis:

- Auxin mimics (2, 4-D, clopyralid, picloram, and triclopyr), which mimic the plant rowth hormone auxin causing uncontrolled and disorganized growth in usceptible plant species;
- Mitosis inhibitors (fosamine), which prevent re-budding in spring and new rowth in summer (also known as dormancy enforcers);
- Photosynthesis inhibitors (hexazinone, bromoxynil), which block specific eactions in photosynthesis leading to cell breakdown;
- Amino acid synthesis inhibitors (glyphosate, imazapyr, and imazapic), which revent the synthesis of amino acids required for construction of proteins;
- Lipid biosynthesis inhibitors (fluazifop-p-butyl and sethoxydim), that prevent the ynthesis of lipids required for growth and maintenance of cell membranes.

Engineering plants for resistance to glyphosate will be discussed. Glyphosate (Nphosphonomethylglycine) is a non-selective, broad spectrum herbicide that is systematically translocated to the meristems of growing plants. It causes shikimate accumulation through inhibition of the chloroplast localized EPSP synthase (5- enolpyruvylshikimate-3-phosphate synthase; EPSPs).

Resistance to the herbicide glyphosate has been developed for soybean. Glyphosate is said to be environmentally-friendly as it does not accumulate in the environment because it is easily broken by soil bacteria. Glyphosate kills plants by preventing the synthesis of certain amino acids produced by plants but not animals. It acts by inhibiting the enzyme 5-enolpyruvyshikimate-3-phosphate synthase (EPSPS), an enzyme in the shikimate pathway and plays an important part in the synthesis of aromatic amino acids in plants and bacteria.

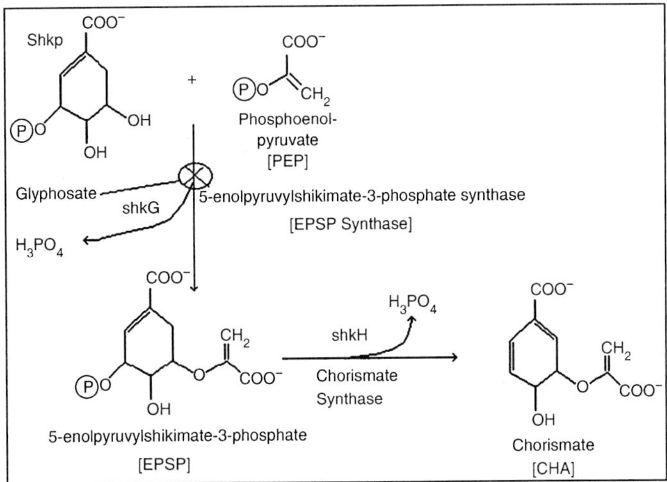

Fig. Mode of Action of Action Glyphosphate

An example of a case where the plant is engineered to inactivate the herbicide before it can act is bromoxynil, a photosynthesis inhibitor. Plants were made resistant to this herbicide by engineering into them the gene for nitrilase obtained from the bacterium, Klebsiella ozaenae. Nitrilase inactivates bromoxynil before it can act.

Engineering Plants for Pathogenic Microbe Resistance

The majority of microbes attacking plants are fungi, but some bacterial diseases of plants do exist. Plants are conventionally sprayed with chemicals to eliminate fungal pathogens. Such chemicals sometimes are not always easily biodegradable, and they may also find their way into food.

A genetic approach which bypasses this problem is to engineer into plants anti-fungal proteins such as the gene coding for chitinase, an enzyme which hydrolyzes chitin, a polymer of the amino sugar N-acetyl glucosamine. Chitinase gene from bean has been cloned into tobacco where chitinase stopped the attack by the fungus, Rhizoctonia solani.

Chitinase is one of the 'pathogen-related proteins' (PRs) synthesized by plants; they also synthesize ant-fungal peptides known as defensins. Genes coding for these are sought from source of high productivity and cloned into plant to protect them.

Plant resistance to bacterial disease has also been genetically engineered. For example, the α-thionin gene from barley has been shown to confer resistance to a bacterial pathogen, Pseudomonas syringae in transgenic tobacco. With regard to engineering plants against viruses, when the viral coat of a plant virus is engineered into a plant, that plant usually becomes resistant to the virus from which the coat comes. Often the plant is also resistant against other unrelated viruses.

Engineering Plants for Insect Resistance

Insect pests are devastating to crops, about US $5 billion are currently being used to control them annually with chemicals. The methods described relate to the use of biological insecticides which are sprayed on plants. Such sprayed insecticides have the disadvantage that thet are inactivated by ultraviolet rays from the sun or may be washed away by the rain. Genetic engineering of crops for resistance against insect pests has the advantage that the active constituents are protected from the environment and remain within the plant The major strategy of producing plants resistant to insect pests is to engineer the gene for producing the toxic crystals of Bacillus thuringiensis (Bt) into plants.

These crystals are produced in Bt but in no other Bacillus sp. They are small proteins and are highly specific against given insects. In such susceptible insects they bind to receptors in the gut lining of the insects, dissolve in the alkali milieu therein and create holes in the gut lining through which gut contents leak out, leading to death. The gene for Bt toxin has been engineered into cotton, tomatoes and numerous other plants.

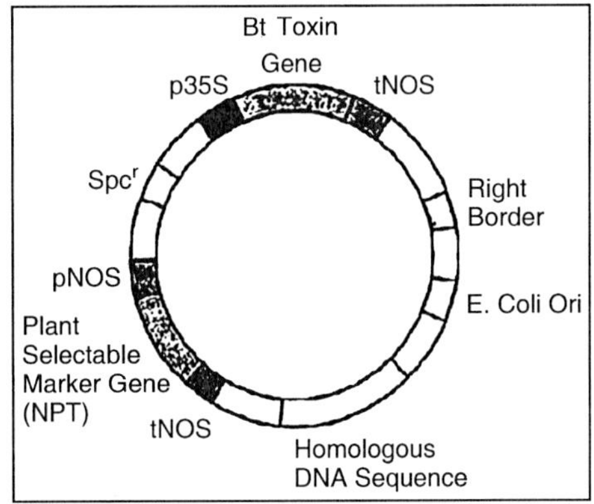

Fig. Cloning Vector Carrying a B. thuringiensis (Bt) Insecticidal Toxin Gene

Alternative strategies which have been inspired by the fact that Bt toxins do not affect some insects, is to engineer into plants two groups of enzymes which inhibit digestive enzymes in the insect gut: amylase inhibitors and protease inhibitors. In effect the insect starves to death. Another strategy for developing insect resistance in plants is to engineer into the plant the gene for cholesterol oxidase, which is present in many bacteria. Cholesterol oxidase catalizes 3-hydroxysteroids to ketosteroids and hydrogen peroxide. Small amounts of this enzyme are very lethal to the larvae of boll weevil which attacks cotton. It is possible that the cholesterol oxidase acts by disrupting the insect larva's alimentary canal epithelium leading to its death.

Genetically Engineering Plants to Survive Water and Salt Stress

Many parts of the world have desert or near desert conditions where water is in short supply. Added to this is the fact that salt used for treating ice in the winter finds its way into agricultural land. These factors create conditions which bring plants into conditions of water (drought) and salt stress. To survive under these conditions, many plants synthesize compounds known as osmoprotectants.

They help the plant increase its water uptake as well as retain the water absorbed. Osmoprotectants include sugars, alcohols and quartenary ammonium compounds. The quartenary ammonium compound, betaine, is a powerful osmoprotectant and the gene encoding it obtained from E. coli has enabled plants into which it was cloned survive drought better than un-engineered plants.

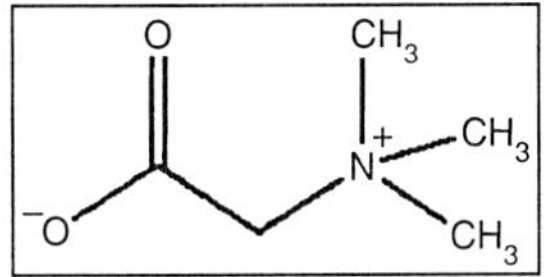

Fig. Betaine

Modification of Plant Consumer Products

This section looks at how genetic engineering has been used to modify the plant food which comes to the consumer.

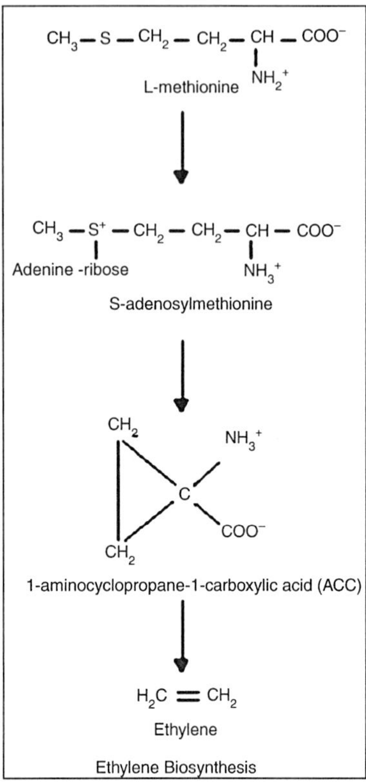

Fig. Synthesis of Ethylene

- *Maintenance of Hardness and Delayed Ripeness in Fruits*: During post-harvest transportation of fruits to supermarkets these fruits sometimes ripen and become soft due to the natural processes which go on within the fruit. These natural processes include the production of polygalacturonase (PG) (which hydrolyzes pectin) and cellulases by the fruit. In tomatoes the softening of the fruit is inhibited by engineering an anti-sense PG producing gene into the plant, enabling the fruit to ripen on the plant before harvesting instead of harvesting them while still green. Such tomatoes have a longer shelf life while retaining the taste of regular tomatoes. The genetically engineered tomato known as Flavr Savr was approved by the FDA in 1994 as safe for human use. In anti-sense technology, a gene sequence is inserted in the opposite direction, so that during transcription, mRNA complimentary to the normal RNA is produced. The anti-sense mRNA therefore binds to the normal inhibiting translation. The net result is that the gene is shut off and in the particular case of PG the fruit-softening enzyme is reduced to about 1 per cent of the normal, thereby inhibiting softening of the fruit and possible microbial attack thereafter. In climacteric fruits (*i.e.*, fruits that are picked before they are ripe) such as tomatoes, avocados, and bananas, the initiation of ripening is associated with a burst in ethylene biosynthesis. After harvesting unripe fruits such as bananas may be treated with ethylene to induce simultaneous ripening. Ethylene has been described as a gaseous plant hormone: extraneous ethylene and ethylene generated by the plant equally induce ripening. Ethylene is a gaseous effector with a very simple structure. In higher plants, ethylene is produced from L-methionine.

- A major step is the production of the non-protein amino acid 1-aminocyclopropane-1-carboxylic acid (ACC), catalysed by the enzyme ACC synthase. It has numerous functions in higher plants. It stimulates the following activities: the release of dormancy, leaf and shoot abscission, leaf and flower senescence, flower opening and fruit ripeneing. Two biotechnological strategies have been pursued to control ethylene action on fruit ripening. One approach taken in tomato was designed to inhibit biosynthesis of ethylene within the plant by the use of antisense expression of ACC synthase. In a second approach, a mutated ethylene receptor from Arabidopsis was introduced into tomato and petunia. This resulted in delayed fruit ripening.

Table. Some Sweet Tasting Proteins Produced by Plants

S/No	Name	Plant	Sweetness Ratio Over Sucrose (w/w)
1	Thaumatin	Thaumatococcus danielli Benth	3, 000
2	Monellin	Dioscoreophyllum cumminsii Diels	3, 000
3	Brazzein	Pentadiplandra brazzeana	2, 000
4	Curculin	Curculingo latifolia	550

- *Engineering Sweetness into Foods*: The taste of fresh tomatoes and lettuce is well known in sandwiches. Some enjoy these items with greater relish with the addition of sweet tasting tomato ketchup. Sweet taste has been engineered into tomatoes and lettuce by cloning into them the synthesized gene coding for monellin. Monellin is a protein which is 3, 000 times sweeter than sucrose by weight; it is naturally obtained from the red berries of the West African plant, Dioscoreophyllum comminsii Diels, and has been expressed in yeast. A major attraction of sweeting tomatoes and lettuce with this protein is that it is 'weight-friendly'. Several sweet proteins which might be similarly engineered into foods are shown in Table.

Table. Modification of Canola oil for Different Purposes

Seed Product	Commercial use(s)
40% Stearic	Margarine, cocoa butter
40% Lauric	Detergents
60% Lauric	Detergents
80% Oleic	Food, lubricants, inks
Petroselinic	Polymers, detergents
"Jojoba" wax	Cosmetics, lubricants
40% Myristate	Detergents, soaps, personal care items
90% Erucic	Polymers, cosmetics, inks, pharmaceuticals
Ricinoleic	Lubricants, plasticizers, cosmetics,
pharmaceuticals	

- *Modification of Starch for Industrial Purposes*: Starch consists of amylose in which the glucose molecules are configured in a straight chain in the α-1-4 linkage, and the branched chain amylopectin which has α-1, 4 and α- 1, 6 linkages. Starches from different plants have different percentages of amylase and amylopectin, but generally in the order of 30 per cent amylase to 70 to 80 per cent amylopectin. Starch is used for making several industrial products such as glue, gelling agent or thickener. For some purposes it may be desirable to have starch that has a preponderance of amylase. When that is the case, antisense technology has been used to block the formation of the amylopectin component of starch, giving rise to a product with only about 20 per cent. One further modification is the engineering into a starch source the enzymes needed to convert starch to high fructose syrup. In the production of high fructose syrup, the starch is first converted to glucose by α-amylase and thereafter the resulting glucose is converted to high fructose syrup by glucose isomerase. Both operations are normally done sequentially. However, both enzymes have been linked together and engineered into potato and the potato starch converted into fructose in one operation with consequent saving in costs
- *Modifying Flower Pigmentation and Delaying Wilting and Abscision in Flowers*: The flower business is of the order of many billions of dollars annually. Most of the market centers around four flowers: roses, carnations, tulips, and chrysanthemums. Hundreds of different flowers differing in shape, size, colour, fragrance, and structure have become available through tradional plant breeding. But the usual shortcomings have also affected this industry: the slow pace of the plant breeding, the uncertainty of the the results of the efforts and the limitation imposed by the paucity of the genes available in traditional plant breeding. Genetic engineering has now been introduced and has helped to extend the range of the variety of flowers. A group of

flavonoids, anthocyanins are commonest pigments in flowers. Anthocyanins are glucosides of phenolic compounds produced in plants, some being colourless, while many are responsible for the colours in plants. The aglycone (non-sugar) protions of anthocyanins are derived from the amino acid phenylalanine. The colour which they bear is determined by the chemical nature of the side chain substituent. By blocking some of the genes in the pathway of anthocyanin synthesis using anti-sense technology or introducing toally new genes it is possible to create flowers with new colours.

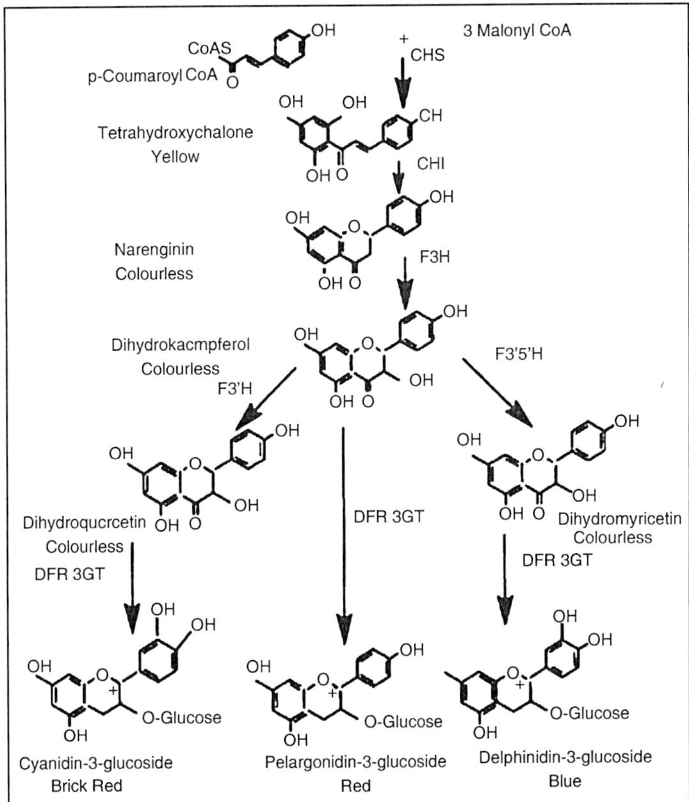

Fig. Synthesis of Anthocyanins in Flowers

It is also known that flower wilting and abscission are controlled by ethylene in the same way as it does with fruits. When a mutated ethylene receptor from Arabidopsis was introduced into petunia, it led to delayed petal fading, and in delayed flower abscission.

- *Modification of Nutritional Capabilities of Crops*: Genetic engineering has enabled the introduction of new nutritional capabilities in crops, in a much shorter time and in a range of qualities impossible with traditional breeding. Unlike genetic engineering which can cross the species barrier, plant breeding deals with the collection of genes within the species. The amino acid content of foods, the lipid composition, the amylose/amylopectin ratio of starch, the vitamin contents and even the mineral contents of foods have all been modified by genetic engineering.

- *Engineering Vitamin A into Rice*: 'Golden rice' has been prepared by engineering it beta-carotene, a substance which the body can convert to Vitamin A to combat vitamin A deficiency (VAD), a condition which afflicts millions of people in developing countries, especially children and pregnant women. Severe Vitamin A deficiency (VAD) can cause partial or total blindness; less severe deficiencies weaken the immune system, increasing the risk of infections such as measles and malaria. Women with VAD are more likely to die during or after childbirth. Each year, it is estimated that VAD causes blindness in 350, 000 preschool age children, and it is implicated in over one million deaths. Golden rice was created by transforming rice with three beta-carotene biosynthesis genes: psy (phytoene synthase) and lyc (lycopene cyclase) both from daffodil (Narcissus pseudonarcissus), and crt1 from the soil bacterium Erwinia uredovora. The psy, lyc, and crt1 genes were transformed into the nuclear genome and placed under the control of an endosperm specific promoter, so that they are only expressed in the endosperm. The plant endogenous enzymes process the lycopene to beta-carotene in the endosperm, giving the rice the distinctive yellow colour which gave it the name 'golden'.
- *Engineering Amino Acids into Legumes and Cereals*: The seed storage compounds in cereals such as corn usually contain proteins deficient in the essential amino acids lysine and methionine. Storage proteins found in many legumes are sometimes deficient in these two essential amino acids, or cysteine. Corn and many grain legumes are used as animal feeds, and feeds made from them have to be supplemented with the deficient amino acids. Legumes such as lupine have been engineered to express sunflower seed albumin which is unusually rich in the sulfurcontaining amino acids methionine and cysteine. High-lysine corn is currently available, but engineering lysine, methionine and or cysteine into corn is almost certainly a matter of time
- *Modifying Fats and Oils for Various Purposes*: Plant oils are derived from soybean, oil palm, sunflower, and rapeseed (canola) and to a lesser extent from the endosperm of corn. Most of the oils is used for margarine manufacture, as fats for baking, for salads and for frying. The extent to which an oil is liquid at room

temperature depends on the degree of unsaturation, *i.e.*, the number of double bonds it has. For industrial purposes oils are also used in cosmetics, in detergents, soaps, confectionaries, and as drying agents in paints and inks. Each use to which the oils are put requires a different property. For example oils which contain conjugated double bonds (in contrast to those which contain double bonds separated by methyl groups–CH2 require less oxgene for polymerization and hence dry more quickly in paints and inks. Genetic engineering has been used to modify oils for various uses. Thus canola oil from rapeseed has been genetically modified for use in various products.

Cojugated Double Bonds:

$$CH_2 = CH - CH = CH - CH_3 \qquad CH_2 = CH - CH_2 - CH = CH_2$$

Soyoil has been genetically modified by DuPont to make it more suitable as an edible oil and also for certain industrial uses including the manufacture of inks, paints, varnishes, resins, plastics, and biodiesel. Soybean oil is a complex mixture of five fatty acids (palmitic, stearic, oleic, linoleic, and linolenic acids) that have vastly differing melting points, oxidative stabilities, and chemical functionalities. The most notable example, developed by researchers at DuPont, is the transgenic production of soybean seeds with oleic acid content of approximately 80 per cent of the total oil.

Conventional soybean oil, by comparison, contains oleic acid at levels of 25 per cent of the total oil. The high oleic acid trait was obtained by down regulating the expression of FAD2 genes that encode the enzyme, which converts the monounsaturated oleic acid to the polyunsaturated linoleic acid. High-oleic oils with elevated oleic acid content are generally considered to be healthier oils than conventional soybean oil, which is an omega-6 or linoleic acid-rich oil. From an industrial perspective, the high content of oleic acid and low content of polyunsaturated fatty acids result in an oil that has high oxidative stability. In addition, soybean oil is naturally rich in the vitamin E antioxidant gamma-tocopherol, which also contributes to the oxidative stability of high oleic acid soybean oil.

High oxidative stability is a critical property for lubricants. Genetic engineering can also be used to produce soybean oil with high levels of linolenic acid, a polyunsaturated fatty acid with low oxidative stability. Soybean seeds with linolenic acid content in excess of 50 per cent of the total oil have been generated by increasing the expression of the FAD3 gene, which encodes the enzyme that converts linoleic acid to linolenic acid The linolenic acid content of conventional soybean oil, in contrast, is approximately 10 per cent of the total oil.

The low oxidative stability associated with high linolenic acid oil is a desirable property for drying oils that are used in coating applications, such as paints, inks, and varnishes. Significant progress has been made in the development of chemical methods for enhancing the functionality of soybean oil for the production of polyols from soybean oil which may eventually lead to a number of industrial applications, including the production of polyurethanes.

Transgenic Animals and Plants as Biological Fermentors (or Bioreactors)

Transgenic animals and plants have been used to produce high-quality pharmaceutical substances or diagnostics. The procedure is known as 'pharming' from a parody of the word pharmaceutical; it is also known as 'molecular farming' or 'gene pharming' and the transgenic plants or animals used are sometimes referred to as animal or plant 'bioreactors' or 'fermentors'. Therapeutically active proteins already on the market are usually produced in bacteria, fungi, or animal cell cultures.

However microorganisms usually produce comparatively simple proteins; furthermore microorganisms are not always able to correctly assemble and fold complex proteins.

If the protein structure is very complicated, such microorganisms may produce defective clumps. In pharming, transgenic animals are mostly used to make human proteins that have medicinal value. The protein encoded by the transgene is secreted into the animal's milk, eggs or blood or even urine, and then collected and purified. Livestock such as cattle, sheep, goats, chickens, rabbits and pigs have already been modified in this way to produce several useful proteins and drugs. Some human proteins that are used as drugs require biological modifications that only the cells of mammals, such as cows, goats and sheep, can provide. For these drugs, production in transgenic animals is a good option.

Using farm animals for drug production has many advantages: they are reproducible, have flexible production, and are easily maintained. Since the mammary gland and milk are not part of the main life support systems of the animal, there is not much risk of harm to the animal making the transgenic protein. To ensure that the protein coded in the transgene is secreted in the milk, the transgene is attached to a promoter which is only active in the mammary gland. Although the transgene is present in every cell of the animal, it is only active where the milk is made.

Some examples of the drugs currently being tested for production in animals are antithrombin III and tissue plasminogen activator used to treat blood clots, erythropoietin for anemia, blood clotting factors VIII and IX for hemophilia, and alpha-1-antitrypsin for emphysema and cystic fibrosis.

A good example of the need for processing a protein in an animal is seen in the silk of the golden spider, Nephila clavipes. The dragline form of spider silk is regarded as the strongest material known; it is five times stronger than steel. People have actually tried starting 'spider farms' to harvest silk, but the spiders are too aggressive and territorial to live close together. They also like to eat each other. Though the genes for dragline silk were isolated several years ago, attempts to produce it in bacterial and mammalian cell culture have failed.

When the genes were put into a goat and expressed in the mammary glands, however, the animal produced silk proteins in its milk that could be spun into a fine thread with all the properties of spider-made silk. This can be used to make lighter, stronger bulletproof vests, thinner thread for surgery and stitches or indestructible clothes. The advantage of using biological fermentors have been put as follows: lower drug prices for consumers, production of drugs unavailable any other way, new valuevalue-added products for farmers, and inexpensive vaccines for the developing world.

- *Lower drug costs*: Expected savings on infrastructure and production costs lead companies producing 'pharm' and industrial crops to predict drug prices 10 to 100 times lower than current prices. The cost of treating a patient with Fabry's disease, currently as much as US $400, 000 a year, for example, is predicted to drop to approximately US $40, 000 annually. Similarly, it is claimed that the leaves from only 26 tobacco plants could make enough glucocerebrosidase, currently one of the most expensive drugs in the world, to treat a patient with Gaucher's disease for a whole year. Regarding plants, the biggest factor in reducing costs is the high yields of recombinant proteins attainable in transgenic plants. Production costs for corn systems are estimated at between US $10 and US $100 per gram for proteins that currently cost as much as US $1, 000 per gram. Dollar figures based on large-scale tobacco production vary widely from less than US $10 per gram to US $1000 per gram. If realised, these projections would represent substantial savings over current costs.
- *Faster, more flexible manufacturing*: Abundantly available commodity like corn and the environment as a production inputs could cut not only production costs but also capital investment in, and the time it takes to increase, manufacturing infrastructure. Very rapid scale-up (or scale-down) of a production pipeline in response to the market or other factors and new drugs could, theoretically, become available sooner.
- *Drugs unavailable any other way*: Cheap production also means that drugs that could not be produced cheaply enough at high volume through conventional methods might become economically viable using genetically engineered crops. Monoclonal antibodies ('plantibodies') fall into this category. One company's idea for such a product is a monoclonal antibody against bacteria responsible for tooth decay, which could be used as a dental prophylactic. A topical therapeutic for herpes, as well as antibodies for the treatment of many other diseases, are also under development.

- *New value-added agricultural products*: These crops producing pharmaceutical products could be a boon to farmers, as they could be economically viable alternatives to commodity production of corn or tobacco.

Special Advantages of Plants as Biological Fermentors

Apart from the advantages accruing in the use of plants and animals as bioreactors, a sector of the biotechnology industry views plants (in comparison with animals) as preferable for protein production for the following reasons.

Table. Human Proteins Synthesized in Animals

Protein	Use	Animal(s)
α-1-anti-protease inhibitor	α-1-antirypsin deficiency	Goat
α-1-antirypsin	Anti-inflammatory	Goat, sheep
Antithrombin III	Sepsis and disseminated intravascular coagulation (DIC)	Goat
Collagen		Cow
Factor VII and IX	Burns, bone fracture, incontinence	Sheep, pig
Fibrinogen	Hemophilia	Pig, sheep
Human fertility hormones	"Fibrin glue, " burns, surgery localized chemotherapeutic drug deliver	Goat, cow
Human hemoglobin	Infertility, contraceptive vaccines	Pig
Human serum albumin	Blood replacement for transfusion	Goat, cow
Lactoferrin	Burns, shock, trauma, surgery	Cow
LAtPA	Bacterial gastrointestinal infection	Goat
Monoclonal antibodies	Venous stasis ulcers	Goat
Tissue plasminogen activator	Colon cancer Heart attacks, deep vein thrombosis, pulomary embolism	Goat

Table. Some Therapeutic Agents Produced in Transgenic Plants

Protein	Plant(s)	Application
Human protein C	Tobacco	Anticoagulant
Human hirudin variant 2	Tobacco, canola, Ethiopian mustard	Anticoagulant
Human gramulocyte-macrophage colony-stimulating factor	Tobacco	Neutropenia
Human erythropoientin	Tobacco	Anemia
Human enkephalins	Thale cress, canola by opiate activity	Antihyperanalgesic
Human epidermal growth factor	Tobacco proliferation	Wound repair/ control of cell
Human α-interferon	Rice, turnip	Hepatitis C and B
Human serum albumin	Potato, tobacco	Liver cirrhosis

Human hemoglobin	Tobacco	Blood substitute
Human homotrimetic collagen I	Tobacco	Collagen synthesis
Human α-1-antitrypsin	Rice	Cystic fibrosis, liver disease, hemorrhage
Human growth hormone	Tobacco healing	Dwarfism, wound
Human aprotinin	Corn surgery	Trypsin inhibitor for transplantation
Angiotension-1-converting enzyme	Tobacco, tomato	Hypertension
α-Tricosanthin	Tobacco	HIV therapy
Glucocerebrosidase	Tobacco	Gaucher disease

- *Plants are less controversial than animal production systems*: Plants do not normally transmit animal pathogens (that can cause diseases like mad cow). Products obtained by 'pharming' plants are promoted as safer than animalsourced proteins. It must, however, be borne in mind that plants as bioreactors may also present their own risks of product contamination from mycotoxins, pesticides, herbicides or endogenous plant secondary metabolites such as nicotine and glycoalkaloids. Using plants as bioreactors also avoids any animal welfare and certain ethical concerns associated with cloning animals and using them as bioreactors
- *Inexpensive, easily delivered vaccines*: Food plants engineered to contain pieces of disease agents can function as orally administered vaccines, avoiding the need for injection and syringes. Currently, tomatoes and other vegetables are under development for that purpose. Although the vaccines would still have to be standardized for dose and delivered to patients one at a time, it is hoped that the lower production costs and the convenience of avoiding refrigeration would make the products attractive to the developing world.

7

The Changing Focus of Industrial Ralations

INTRODUCTION

Industrial relations in countries, sub-regions and regions, have been influenced by a variety of circumstances and actors such as political philosophies, economic imperatives, the role of the State in determining the direction of economic and social development, the influence of unions and the business community, as well as the legacies of colonial governments. Over several decades IR in many industrialized market economies of the West, and also in Australia and New Zealand in the Asia-Pacific as well as in the South Asian countries, paid less attention to competitiveness than did the younger 'discipline' human resource management.

IR fulfilled the function of providing employees with a collective voice, and unions with the means to establish standardized terms and conditions of employment not only within an enterprise but also across an industry, and sometimes across an economy. This was achieved through the freedom of association, collective bargaining and the right to strike. Similar results were achieved in the South Asian sub-region where political democracy, and sometimes socialist ideology, provided enormous bargaining power and influence on legislative outcomes to even unions with relatively few members. A different IR regime emerged in some of the South-East and East Asian economies, driven by competition in export markets and different political systems bearing little resemblance to the values underpinning Western-style democracies.

While we will return to this subject, it is worth noting that during the past decades labour relations were often viewed by Asian governments as a means of minimizing conflict, preventing union agitation, or as in the case of India and Sri Lanka, of controlling employers and winning votes. Conflict resolution was achieved through dispute prevention and settlement mechanisms external to the enterprise, such as conciliation, arbitration and labour courts. In South Asia the objective was also achieved through restrictions and prohibitions on the freedom of action of employers in matters such as termination of employment, closures and even transfers of employees. On the other hand, several South East Asian countries resorted to measures to restrict trade union action and to control unions, as well as to avoid union multiplicity.

In South Asia while the focus of IR was on equity from the point of view of workers and unions, in South-East Asia the emphasis was on economic efficiency and less on worker protection laws. Low unionization in many Asian countries, strong governments in South-East Asian countries and the Republic of Korea, and perceptions that unions can be potential obstacles to a particular direction of economic development, led to a relative neglect of IR. Moreoever, hierarchical management systems and respect for authority, which have mirrored the external social system, have been inconsistent with consultation, two-way communication, and even with the concept of negotiating the employment relationship.

Japan, however, was an exception where, since the 1960s, workplace relations and flexibility facilitated by enterprise unionism dominated IR in the larger enterprises. Australia and New Zealand have traditionally focused on centralized IR, though the emphasis has radically changed in New Zealand during this decade, and is changing in Australia.

Globalization has led employers to push for less regulation of IR, less standardization of the employment relationship, and a greater focus on the workplace as the centre of gravity of IR. Employers as well as some governments are viewing IR from a more strategic perspective, i.e., how IR can contribute to and promote workplace cooperation, flexibility, productivity and competitiveness. It is increasingly recognized that how people are managed impacts on an enterprise's productivity and on the quality of goods and services, labour costs, the quality of the workforce and its motivation. The industrializing Asian countries which recognize limited labour rights are gradually coming to terms with union pluralism and agitation, and the need for less hierarchical and paternalistic approaches to managing people.

On the other hand, traditional IR, which developed in the context of mass manufacturing, is viewed by many employers as less appropriate to the growing service sector, the emerging knowledge workers, and the proliferation of work performed outside the enterprise, under arrangements which do not fit the traditional IR concepts of standardized employment terms and conditions for people working within an enterprise.

The diminishing role of the State as employer (still important as it is in several countries) may correspondingly reduce the State's interest in intervening in IR. The direction in which economies in transition is heading is uncertain, but indications are sufficient to raise the question whether over-regulation of the labour market through legislative activity is a possibility, in which event it will stifle the flexibility employers will need in time to come to compete in the global marketplace in more value-added industries than the ones they are engaged in at present.

Employers and employers' organizations need to influence the industrial relations system (including the labour law) in the context of competitiveness. Not all employers' organizations in Asia-Pacific are adequately equipped to do so, so that they have t©acquire the requisite knowledge base needed to influence the policy environment. This also implies that employers' organizations will have to develop a strategic perspective of IR, in the same way that employers are seeking to develop HRM policies and practices which foster competitiveness. In this task it is inadequate to merely espouse the familiar claim that labour markets should be deregulated. It is sometimes not clear whether this claim means that we should dispense with labour laws, or whether it means that there should be a careful identification of the legitimate areas of legal prescription.

If it is the former, it is difficult to see how there can be social stability if one group is to be granted unrestricted freedom of action. If it is the latter, it means being able to identify the interventions which obstruct the efficient functioning of the market. Efficiency (on behalf of which deregulation is espoused) and equity are not antithetic concepts. Rather, it is efficiency and inequity which are antithetic as inequity leads to inefficiency. What we need to do is to develop a perspective of labour law and IR which is based on the premise that it is not economies which compete, but enterprises and clusters of competitive industries, a theme we will return to.

Policy makers and unions also need to address the issue of IR in the context of competitiveness. If they do not, it will further compel employers to resort to HRM as the more relevant means for achieving corporate objectives. It has to be appreciated that IR did not grow out of a need to develop competitive strategies, while this has been precisely the background and impetus to the development of HRM, even if effective HRM in practise still remains islands of excellence. With the declining importance of collective IR, the increasing interest in workplace relations and the weakening of unions in many industrialized countries, IR will have to accommodate employment relations in the non-union sector and the individual employment relationship. As such, it may even need to change its name or label to 'industrial and employment relations' to more accurately reflect the reality.

Despite the shift of emphasis away from collective IR in the 1990s, there is a growing interest in Asia in the IR systems of other countries, both within and outside the region. This is most evident in economies in transition, as we shall see. These governments, as well as representatives of employers and employees, are looking for models from other countries which can be adapted to their own conditions. This search is prompted because of the need to establish an IR system relevant to the emerging business environment. It is also a result of the recognition of the necessity to develop an IR system which contributes to social stability which can otherwise be eroded by disputes and conflicts. This search is not confined to economies in transition. In Thailand, for instance, the labour administration authorities have commenced, with the support of the ILO, a project aimed at encouraging employers and workers to establish better workplace relations and mechanisms. Here again they are seeking to identify successful models and experiences for purposes of benchmarking and adaptation.

WORKPLACE RELATIONS

Early Emphasis and Current Trends

At its inception the labour market was dominated by the classical economics view which espoused free and unregulated labour markets. This laissez-faire capitalism led to social injustices and inequities since labour did not have the power to bargain with employers in terms in a way which even approached a degree of equality in bargaining strength. Additionally, the dominant position of the employer in what was formerly termed the "master and servant" relationship prevented labour from enjoying rights.

IR therefore came to espouse a degree of labour market regulation to correct the unequal bargainig power. It was natural that IR developed in the context of the theory that problems in labour relations emanate largely from market imperfections which operate against the interests of labour and cause imbalances in the power relationships between employers and employees. The causes of labour problems - even those within the enterprise - were thought to need addressinh through a range of initiatives external to the enterprise, by

- The State through protective labour laws and dispute settlement mechanisms
- Voluntary action on the aprt of employees to protect themselves and increase their bargaining strength through freedom of association and collective bargaining, but backed by State interventions to guarantee these rights.

The focus on relations external to the enterprise - especially through national/industry level collective bargaining - was initially welcomed to an extent even by employers in several industrialized countries because it reduced (through standarized terms) competitive advantage based on labour costs. Besides, it took the most contentious issue - wages - out of their direct responsibility and transferred its negotiation to the representatives of the social partners, viz., unions and employers' organizations. Unions naturally welcomed it as it gave them an influential base outside the enterprise by uniting employees from many organizations on either an industry or national basis. It also provided an important role in IR for employers' organizations as well as for the State.

When 'transferred' to developing countries, it sometimes had disastrous consequences because it facilitated the multiplicity and politicization of unions. In this regard, the workability of an IR system in which the emphasis was on major decisions taken outside the workplace presupposed a high rate of literacy, as well as of education and awareness among employees, and an ability to monitor the actions of their representatives operating at a level far removed from the workplace. This was lacking in the case of developing countries. In these circumstances for the greater part of this century the work of IR academics and researchers largely focused on trade unionism, collective bargaining, labour law, dispute settlement and, generally, on the external environment. Their interest in what occurs at the workplace level is of very recent origin.

The emphasis on collective IR was facilitated by government intervention in IR through legal prescriptions (which emphasized the normative aspects of IR), as well as by the growth of union power which depended on decions to be implemented at the enterprise level being made outside the enterprise in a way which, like laws, had a normative effect. Consequently, in many countries rules and obligations on employers were imposed from outside in respect of an essentially bipartite relationship at the enterprise level. Therefore organizations had limited flexibility to effect changes, which had to be made within the parameters set by the the externally imposed norms. Among Western countries enterprises in the USA have been the least hampered by external norms in making adjustments at the enterprise level to respond to changes.

Collective IR operates in three ways. One way is through national or industry level agreements between unions and employers' organizations. A second way is through agreements between a single employer and a union. A third way is through legislative enactments applicable to employers and employees generally, or to particular sectors, or to particular categories of employees. National agreements may sometimes stipulate general principles, leaving the details to be worked out through negotiation at the enterprise level.

Regulating the internal labour market (the enterprise) through means external to it redressed the balance of power between labour and management and, by settling terms and conditions of employment, reduced conflict at the enterprise level on such matters. A more 'enterprise-based' view - long ignored by most academics and recognized by employers during the last two decades - saw that labour problems or issues are due not only to conflict over terms and conditions of employment usually negotiated between unions and employers. They are reflected in or emanate from, for example, low productivity, absenteeism, high labour turnover, lack of job security, the physical work environment, repetitive jobs, lack of motivation, the failure to recognize the performance of individuals or groups inherent in standardized wage systems, and lack of training.

Many of these problems or their causes required them to be addressed at the enterprise level. Hence the view emanating from HRM and increasingly important since the 1980s, that labour problems arise not so much from factors external to the enterprise, as from the unsatisfactory management of human resources within the enterprise. Hence the corrective actions needed to be taken at the enterprise level, with the objective of ensuring as far as possible (through appropriate HRM policies and practices) a convergence of organizational and individual goals and needs, or a reasonable balance between them. Other developments which challenged the dominance of collective IR were the employers' claim to flexibility needed to adapt to changes required to compete in the global marketplace, the emergence of atypical forms of work alien to the standardization sought to be achieved through collective IR, and the changing character of the workforce.

IR problems need to be addressed by means which are both internal and external to the enterprise. Some basic conditions of employment cannot be left to each individual employer to decide. The problem is that there has been over-emphasis on external regulation, and IR has lost sight of the fact that in the final analysis the quality of an IR system has to be judged by how it works in practise - and that the centre of gravity of IR is the workplace. It also lost sight of the fact that sound relations have to be built from within the organization. The lacuna thus came to be filled by HRM, and employers pushed for moving negotiations to the workplace level.

In developing countries the situation has differed from country to country or region to region. In many of them as in Asia, negotiations outside the workplace leading to normative agreements covering large sectors of the economy have been the exception. This has been due to

- Low unionization rates
- Control of unions by governments in some countries
- State prescription of the rules of the IR system, reducing the scope for negotiated agreements having a normative effect. This has been mainly due to two reasons. In the more 'union friendly' South Asian region it has been the result of the political patronage extended to unions who have used their influence to secure through legislation what they could not through negotiation. In the less 'union friendly' South East and East Asian region, it has stemmed from the desire to control unions and to ensure that they do not seek to disrupt the direction of economic development.

In more recent times IR has been influenced by other social sciences such as organizational psychology and behaviour. Traditionally economics and law were the two main influences on IR, which led to a concentration on macro level IR, and therefore on unions, government and collective bargaining, important as they were. Organizational behaviour has been influenced by psychology which centres on the individual, and by social psychology which focuses on relationships between people and on group behaviour. It is easy to see, therefore, why HRM has been influenced by the other social sciences. Paradoxically IR, though dealing with 'relations', has largely ignored until recently the social sciences relevant to human behaviour within organizations. The reason is that the 'relations' it concentrated on were relations external to the organization, including relations with the State.

Reasons for Emphasis on Workplace Relations

Several changes in recent years have been responsible for more attention being paid to employment relations within organizations. The first is the impact of globalization which has significantly changed the ways in which enterprises are managed and work performed. Enterprises have resorted to a range of measures to increase efficiency and competitiveness, based not on low wages and natural resources, but on innovation, skills and productivity as ways of improving quality and reducing costs.

Since productivity and quality have become major considerations in competitiveness, the quality of the workforce and training have become critical factors. Shorter product life has enhanced the need for multi-skilled easily trainable employees. Emplyee skills have become important determinants not only of flexibility, productivity and quality, but also of employability, investment and the ability to rapidly adapt to market changes.

A second development which has shifted attention to workplace relations is technology. On the one hand, technology management is possible only through people, and the way they are managed and trained affects the success of such transfer. Technology is also displacing traditional jobs and creating new jobs requiring different skills. Further information technology, the limits of which are not known in terms of its potential to effect change, is exerting a tremendous impact on the structure of organizations, the nature and location of work and the way it is organized. In socieites of the future information and knowledge will be - as in fact they already are - crucial to competitiveness.

Technology is already facilitating changes in organizational structures creating flatter organizations. This has resulted in management effected less by command and supervision, and more through emphasis on cooperation, information-sharing and communication, and with a more participative approach to managing people. Modern technology now makes it possible for aspects of work to be performed outside the enterprise, for example from home, and even outside national borders. Part-time work is increasing particularly due to the influx of more females into employment and their preference in some cases for part-time work. Developing countries are also feeling the impact of these changes.

A third factor is the changes occurring in workforces, to varying degrees, in both industrialized market economies and developing economies. Many countries have witnessed the emergence of workforces with higher levels of education and skills than before which need to be managed in a manner different from the way in which employees, especially blue collar employees, have hitherto been managed. This will assume more importance in the future as a result of the enlarging service sector and the growth of knowledge-intensive industries. The skills of an employee are, therefore, an issue on which the interests of employers and employees converge, and the "development" of the employee is now of greater mutual advantage to employers and employees. Hence the greater need than before for cooperative and participative forms of IR.

Further, the many emerging work arrangements do not fit into the traditional employment relationships. Increasing numbers of enterprises are differentiating between the core and peripheral workforce, which consists of those whose work can be performed by persons outside the enterprise who specialize in it. The tendency is to contract with outsiders to perform this work. Even manufacturing companies are becoming essentially assembly firms, and many service organizations now act as brokers "connecting the customer with a supplier with some intervening advice." The increasing number of temporary and part-time employees in the rapidly expanding service industries, some of which experience peak periods (hotels, airlines, shops) require a flexible labour force.

Thus, instead of one workforce, we are moving towards these various groups, each with different contractual arrangements and requiring to be managed differently. The indications are that at the beginning of the next century less than half the workforce in industrialized countries will be in full-time employment within the enterprise as we know it. These trends will not be confined to the highly industrialized countries, but will appear in the fast growing economies as costs rise, competitiveness increases, and more women participate in economic activity. In summary, traditional (collective) IR (as did management) evolved in an age of mass production systems. Among its characteristics are the following:

- An emphasis on solving IR problems through means external to the enterprise.
- Standardization of employment terms which was appropriate to the largely repetitive tasks, narrow job classifications that prevailed, and to less educated workforces as well as hierarchical management structures. The notion of standardization - acceptable to employers as well - meant that the employment relationship could be standardized either through external means (laws and national/ industry agreements), or through agreements at the enterprise level prescribing standard terms and conditions of employment.
- Coverage of employees by standard contracts providing for fixed hours, remuneration fixed by law or by collectively bargained agreements.
- IR developed at a time when services were less important to the economy than they are today, and in an age when most people (other than those in agriculture) physically worked within an enterprise. The idea of people working for an organization but not necessarily in an organization is of recent origin.

Globalization and Industrial Relations

The pressures on traditional IR models are not all due to globalization, as we shall see, but many of the changes taking place can be traced to globalization. it is not always easy to disentangle the causes and effects of globalization. However, it would probably be true to say that globalization is represented by the opening up of markets due, in large measure, to foreign direct investment consequent upon the lowering of investment barriers in practically all countries; by the liberalization of trade, and by the deregulation of financial markets in consequence of which governments increasingly have little control over the flow of capital across borders.

All this implies the dominance of the market system, facilitated by the collapse of alternative economic (and in many cases political) systems. There is also a direct link between globalization and information technology (IT). Rapid technological change and reduction in communication costs have facilitated the globalization of production and financial markets. At the same time globalization stimulates technology through increased competition; it diffuses technology through foreign direct investment. As aptly observed:

"Together, globalization and IT crush time and space."

These developments have had further effects such as:

- Democratization and pressures for more labour rights in countries where such rights have been restricted
- More liberalization and deregulation
- Competition for investment
- Increased economic independence of nations
- Capital, information and technology flows are on the increase
- Internationalization of enterprises and creation of mergers and alliances
- Customer-driven (and not product-driven) global and local markets, but at the same time segmented markets
- Competitiveness increasingly based (not on low wages or natural resources) on knowledge/innovation, skills and productivity. The success of global companies is to a large extent dependent on their ability to organize (within and between organizations) across national boundaries information, money, people and other resources.

Employer Responses and Implications for Industrial Relations

Among the responses of employers are the following:

- Moving production overseas to reduce costs and to facilitate sensitivity to local and regional market requirements.
- Contracting out and out-sourcing. It is an important rationale of out-sourcing that it, on the one hand, enables an enterprise to concentrate

on its core competencies, and on the other hand, it makes service work more productive. For example, in the USA, outsourcing of functions in hospitals not directly related to the work of doctors and nurses (care of patients) has substantially increased the productivity of the hospitals, and provided new opportunities for service employees. "Outsourcing is needed not just because of the economics involved. It is required equally because it gives opportunities, income and dignity to service work and service workers."

- More part-time and temporary work (especially among women, the elderly and students)
- Introduction of new technology
- Pushing for a more deregulated and flexible labour market
- More emphasis on productivity and quality
- Greater employee involvement in the design and execution of work
- Shifting the focus of collective bargaining from the nation/industry level to the enterprise level. Employers are of the view that issues relevant to the employment relationship such as work re-organization, flexible working hours and contractual arrangements, and pay for performance and skills, are increasingly workplace-related, and should therefore be addressed at the enterprise level. In the USA collective bargaining has, with some exceptions, been very much at the enterprise level; in the UK there is a marked shift towards enterprise bargaining; and the trends in Continental Europe are also in that direction. In many Asian countries outside Australia and New Zealand, the relatively little collective bargaining has been mostly at the enterprise level. In New Zealand negotiation has in the 1990s been almost entirely decentralized, and in Australia the trend is in the direction of decentralization.
- Exceptionally (in the USA) employers have reduced terms of employment through 'concession bargaining' when firms have been in financial difficulties.
- Downsizing the workforce.

One important response has been the introduction of flexibility in the employment relationship to increase the capacity of enterprises to adapt rapidly to market changes. This has involved measures such as

- Flexible working hours
- Part-time work
- Different types of employment contracts to the standard ones familiar to collective IR
- Flexibility in functions, so that employees who are multi-skilled are not confined to the performance of only one task. They can cover up for absenteeism, and make some jobs redundant.

- Flexible pay which involves some component of pay being dependent on performance, whether of the company, a group or the individual.

Globalization has, through technology diffusion, substantially increased the introduction of new technology. This, as well as the need for flexible adaptation to market changes, have led to the re-organization of production systems and methods of work, such as the following:

- Reduction of narrow job classifications and demarcation lines between managers and workers, accompanied by skills enhancement needed to perform jobs with a broader range of tasks.
- Increasing areas for worker involvement in the conception, execution and control of work.
- A greater focus on workplace relations and policies and practices conducive to better motivation and performance such as information-sharing and two-way communication.

These responses have increased the necessity for employers to make more investments in skills training, to offer incentives to employees to improve their skills, and for workers to take upon themselves some responsibility for their own development. The competition generated by globalization and rapid technological changes accompanied by shorter product life have, while destroying countless jobs in industrialized countries, created opportunities for multi-skilled and easily trainable workers, and for the most significant group of emerging employees - the knowledge worker. Knowledge and skills have become the most important determinants of investment, employment opportunities, productivity and quality and of flexibility.

The impact globalization and information technology have had on each other has made work more mobile, capable of being performed in different parts of the world without the need to actually set up physical facilities in other countries. Other changes in the nature of work and workers are being brought about partly by globalization, but not entirely because of it. For instance, it is arguable whether globalization is solely responsible for the growing service sector, and it does not account for the rapid influx of women into the workforce. Be that as it may, some of the changes which have a fundamental impact on traditional IR include the following:

- The expanding service sector at the expense of the manufacturing sector in industrialized and rapidly industrializing countries
- More advanced and skilled workforces
- The rapid influx into the workforce of women who will, in some countries, occupy more than half the emerging jobs
- An increasing number of people who will not be working in an organization, though they will be working for an organization
- The decreasing number of people working under 'permanent' contracts of employment, and the proliferation of other types of work

arrangements such as part-time and temporary work, home work and contract work. Thus traditional IR has been challenged to accommodate different types of employment contracts, and different types of pay systems to reward performance and skills.

Competitive Advantage and Industrial Relations

The push for competitive advantage render it necessary for IR, without abandoning its earlier function of providing mechanisms to secure a fair distribution of the gains of economic performance, to also develop a strategic perspectivc which promotes the goal of competitiveness. This implies an IR system not only from the point of view of bargaining, conflict resolution, basic rights and so on, but also from the point of view of contributing to enterprise performance on which the welfare of employees also depends.

Viewing IR from the perspective of competitive advantage involves distinguishing between competitive and comparative advantage. Whether it is economies or enterprises that truly compete in the global marketplace is a complex issue. But certainly the overall national environment of which IR is a part, contributes to the competitiveness of industries and enterprises. However, as pointed out by Michael Porter, there is "no accepted definition of competitiveness. To firms competitiveness (means) the ability to compete in world markets with a global strategy."On the other hand, to some politicians it signifies a positive balance of trade, while to some economists it means a low unit cost of labour adjusted for exchange rates.

Comparative advantage of a country is based on the greatest advantage (or the least disadvantage) a nation has in relation to its trading partners. Where this advantage depends on a natural resource (e.g. oil) and wages rise to a level similar to that of competitors, the comparative advantage is lost. Dependence on low skilled low cost labour renders it impossible to add value to the particular advantage or resource. Where the investment in education and skills development has been low or inappropriate, moving to more value-added activities becomes difficult, if not impossible.

Competitive advantage, by contrast, is based on the ability to add value to the advantage or resource. High and appropriate investment in education and skills enables the absorption of higher wages, which are compensated or off-set by higher productivity/quality and the capacity to innovate and introduce technological change. Today, of the many factors on which competitiveness is based,people are a key factor. Their contribution to competitiveness is conditioned by their literacy, type and quality of education, their skills, motivation, work attitudes, value systems and quality of life.

If, therefore, employers do not see IR as contributing to strategic management objectives, they will continue to view the techniques of HRM as being more relevant to their objectives, as a way of individualizing the employment relationship and keeping decision-making within the enterprise. Hence the issue - to be covered elsewhere in this Paper - of harmonizing IR and HRM.

Management and Industrial Relationsn

These developments impact not only on IR. Technology, information and the imperatives of succeeding in the global marketplace are transforming the way enterprises are managed as well as management objectives themselves. The much closer relationship between management and IR needs to be recognized - principally because both must focus on people and IR at the workplace level.

Industrial relations systems and practices are shaped by the three main actors - governments, workers/unions and employers/employers' organizations. In Asia, for instance, governments have had a significant impact on IR. In Continental Europe unions in some countries (Germany and the Nordic countries) have had a major influence on shaping IR. In the UK the Thatcher governments had a significant influence in changing the direction of IR. In the USA it has been more employer-driven. However, it has been too infrequently appreciated by IR specialists that IR are shaped to a large extent by the way enterprises are managed, and that when fundamental changes occur in management they create changes in IR as well.

This is particularly so in the context of globalization; many of the changes taking place in IR - indeed the increasing shift from collective IR to more enterprise (and in some cases, individual) focused IR - are spurred by employers. It is significant that the collective standardized model of IR evolved during the era of the classical model of the enterprise which itself sought standardization at the expense of creativity and innovation.

Therefore in this context it is useful to note two important - and diametrically opposite - theories about management. The first and earlier theory is to be found in the scientific management school, which viewed the worker as a mere cog in the organizational structure. Since, according to Taylor, the worker does not possess creative ability let alone intelligence and wisdom, the elements of a human-oriented management system which promotes sound industrial relations such as communication, consultation and participation, found no place in the theory.

The hallmarks of organizations based on this model are centralized and clear lines of authority, a high degree of specialization, a distinct division of labour, numerous rules pertaining to authority and responsibility, and close supervision. This concept of management can be seen as an ideal breeding ground for an industrial relations system based on conflict rather than on cooperation. The opposite theory, appropriately styled the human relations school, had as one of its earliest and celebrated exponents, Douglas McGregor. He gave an impetus to the development of a management theory which focused on the human being as part of an enterprise, which, in turn, was viewed as a biological system, rather than as a machine. Human relations, trust, delegation of authority, etc. were some of the features of this theory. In the preface to his classic.

The Human Side of Enterprise McGregor underlined the necessity to learn "about the utilization of talent, about the creation of an organizational climate conducive to human growth ... This volume is an attempt to substantiate the thesis that the human side of enterprise is 'all of a piece' - that the theoretical assumptions management holds about controlling its human resources determine the whole character of the enterprise. They determine also the quality of its successive generations of management."

Two basic realities of an organization in McGregor's model is the dependence of every manager on people under him and the potential of people to be developed to match organizational goals. He therefore postulated that people are not by nature resistant to change in an organization, and that people have the potential to be developed and to shoulder responsibility. As such, management's main task is to organize business in such a way as to match people's goals with organizational ones. McGregor believed that the dynamism for organizational growth is found in the employees of the organization. It could be said that in McGregor's Theory Y (as it is called) is to be found the essence of human-oriented management and workplace industrial relations systems. The events noted in the succeeding paragraphs which are compelling enterprises to pay greater attention to the human factor in management, serve to vindicate McGregor's basic theory propounded as far back as 1960, if not earlier.

However, subject to exceptions (such as Japan in Asia) most large enterprises continued to be dominated by hierarchies. This is reflected in the classic "strategy, structure, systems" (the three Ss) of modern corporations, vividly expressed by two writers:

“Structure follows strategy. And systems support structure. Few aphorisms have prevented western business thinking as deeply as these two. Not only do they influence the architecture of today's largest corporations but they also define the role that top corporate managers play."

As explained by Bartlett and Ghoshal, in this concept of an enterprise top level managers see themselves as the designers of strategy, the architects of structure, and the managers of systems. The impact of the three Ss was to create a management system which minimized the idiosyncracies of human behaviour, emphasized discipline, focus and control, and led to the view that people were "replaceable parts".

The basic flaw - particularly in the context of today's globalized environment - of this concept is that it stifled the most scarce resource available to an enterprise: the knowledge, creativity and skills of people. Successful enterprises have now moved away from this corporate design, and their philosophy, which has transformed corporations enabling them to compete in the new competitive environment, consists of the following:

"First, they place less emphasis on following a clear strategic plan than on building a rich, engaging corporate purpose. Next, they focus less on formal structural design and more on effective management processes.

Finally, they are less concerned with controlling employees' behaviour than with developing their capabilities and broadening their perspectives. In sum, they have moved beyond the old doctrine of strategy, structure, and systems to a softer, more organic model built on the development of purpose, process and people."

Those enterprises which have effected a successful transformation to a more 'people-focused' organization recognize that the information necessary to formulate strategy is with their frontline people who know what is actually going on, whether it be in the marketplace or on the shopfloor. The chief executive officer, for instance, can no longer be the chief architect of strategy without the involvement of those much lower down in the hierarchy.

How do these developments relate to enterprise level labour relations? In essence, they heighten the importance of the basic concepts of information sharing, consultation and two-way communication.

The effectiveness of the procedures and systems which are established for better information flow, understanding and, where possible, consensus-building, is critical today to the successful managing of enterprises and for achieving competitiveness. As such, the basic ingredients of sound enterprise level labour relations are inseparable from some of the essentials for managing an enterprise in today's globalized environment. These developments have had an impact on ways of motivating workers, and on the hierarchy of organizations. They are reducing layers of management thus facilitating improved communication.

Management today is more an activity rather than a badge of status or class within an organization, and this change provides it with a wider professional base. The present trend in labour relations and human resource management is to place greater emphasis on employee involvement, harmonious employer-employee relations and on practices which promote them. One of the important consequences of globalization and intense competition has been the pressure on enterprises to be flexible. Enterprises have sought to achieve this in two ways. First, through technology and a much wider worker skills base than before in order to enhance capacity to adapt to market changes. Second, by introducing a range of employee involvement schemes with a view to increasing labour-management cooperation at the shop floor level, necessary to achieve product and process innovation.

Achieving flexibility does not depend on the absence of unions. Organizational flexibility "depends upon trust between labour and management.

It implies that workers are willing to forego efforts to establish and enforce individually or through collective action susbtantive work rules that fix the allocation of work, transfer among jobs, and workloads. Organizational flexibility also implies that workers are willing to disclose their proprietary knowledge in order to increase labour productivity and the firm's capacity for innovation."

Manufacturers in Japan and Germany were more successful than those in, for example, Britain and France in achieving this flexibility. The two former countries after World War II institutionalized labour-management consultation. Many organizations have successfully transformed themselves to promote the role of trust. In strategies involving the use of technology and promotion of innovation, employees are a critical factor. The requirement of organizational flexibility and its industrial relations and human resource implications have had a major impact on the way organizations are structured (less hierarchical), how authority within the firm is exercised (less unilateral), and on how decisions are arrived at and work organized (through information sharing and consultation, transfer of more responsibility to employees and cooperative methods such as team work).

Traditional assumptions that efficiency is achieved through managerial control, technology and allocation of resources have given way to the view that efficiency is the result of greater involvement of employees in their jobs, teams and the enterprise. Organizations which have made this shift tend to reflect the following characteristics: few hierarchical levels; wide spands of control; continuous staff development; self managing work teams; job rotation; commitment to quality; information sharing; pay systems which cater to performance rewards and not only payment for the job; generation of high performance expectations; a common corporate vision; and participative leadership styles. It hardly requires emphasis that achieving most of these requires training. In Asia too there is a keen awareness in the business community that radical changes are necessary to sustain Asia's dynamic growth. The point is well made by a noted Asian:

The earlier generation's recipe for success hinged on hard work, smart moves, the right business and political connections, monopolies, protectionist barriers, subsidies, access to cheap funds and, in many cases, autocratic leadership and a docile labour force.

The 'global village' is this system's nemesis ... the new 'Global-Asian' manager has to exercise greater levels of leadership than before, and balance this with being an entrepreneur, modern manager and deal-maker skilled at public relations. To this has to be added coaching, team-building and motivating the company, the ability to visualize, plan strategically, market and re-engineer products and services, and the belief in a customer driven culture.

None of these shifts is feasible without a substantial change in traditional models of dealing with people in an enterprise.

The Multinational Dimension

Globalization has substantially increased the influence of multinationals (MNEs) on the IR environment. A few figures indicate their importance in the global economy:

- From about 7,000 MNEs in 1969, the number has increased to over 37,000 today
- Though employing directly less than 2 per cent of the world's labour force, they account for one-third of the world's private sector assets, and have world-wide sales of US $ 5.5 trillion, which is marginally less than the USA's GDP in 1993.
- In 1995 foreign direct investment was in the region of US $318 billion, with the USA receiving US $60 billion and the UK US $30 billion.

What is significant about the increased presence of MNEs, especially in developing countries, is that governments are competing to attract them and are prepared to change the environment (including the IR environment) to achieve this objective. This gives the MNEs the opportunity to spur changes which have commenced - for instance, in economies in trasnition. The point is graphically made by the well-known former CEO of Asean Brown Boveri, Percy Barnevik:

Global companies speed up the adjustment. We don't create the process, but we push it. We make visible the invisible hand of global competition.

Some Influences on Asian Industrial Relations

Main Influences

The IR systems in Asian countries have emerged from circumstances and values somewhat different to those which underpin and which have shaped Western IR systems. Western models of IR do not adequately explain, and help us to understand, the shape that Asian IR have assumed.

This is so despite the fact that the labour laws of several developing countries (including Asian ones) have, to a greater or lesser extent, been influenced by Western countries, whether or not as a result of colonization. The main features of developing country industrial relations systems which distinguish them from those of the industrialized West have been well explained as follows:

".... a dualistic economic structure, where a pre-capitalist economic system mainly dominates the scene; a small industrial sector and the related small numerical size of the working class; a segmented labour market, where a sharp dualism both between modern and traditional manufacturing sectors and small and large firms exists; the dominance of the state in the industrial sector; weak trade unions, and thus the absence of collective bargaining between employers and employees."

About thirty five years ago a group of famous writers claimed in a seminal and celebrated work that industrialization, which would occur in all countries despite their different stages of economic development and cultures, will result in countries having similar systems. The concept of the universalism of industrialization and pluralistic industrialism based on the idea of convergence between the West and developing countries achieved through the unifying influence of science and technology never came to pass. The reasons for this explain the current shape of IR in Asia and many Asian views on IR.

One fundamental reason for the divergence between Western and Eastern (and developing country) IR systems is to be found in the different industrialization processes followed, and the consequences for social systems such as IR. On the whole, Western industrialization did not take place under State direction or patronage, but in a laissez-faire setting in which an entrepreneurial middle class moved the industrialization process forward, which in turn created a distinctive working class (proletariat).

Due to its relative homogeneity, this working class found it possible to organize themselves collectively into trade unions to protect and further their interests. Western governments did not, unlike some developing ones, 'create' unions.

Western IR systems reached maturity in this century, long after the commencement of the industrial revolution and at a time when the current pluralistic and democratic political systems were more or less in place. The systems which emerged - however different they were from one country to another - had certain essential features: they were underpinned by a value system based on pluralism and democracy, a balance between employers and employees, and relatively minimal government intervention in the industrialization process and in IR. In such an environment collective bargaining - a fundamental institution of Western IR systems - and freedom of association - a fundamental philosophy underlying such systems - were logical developments.

On the other hand, the majority of Asian and other developing countries were subject to foreign occupation during which period no indigenous entrepreneurial middle class of any significance emerged which could have spearheaded the industrialization process in the post-colonial period. During the colonial period governments assumed a dominant role, one they did not have in their home countries. This role was maintained by the post-independence governments, albeit with different results. It was only after the industrialization process had been in operation for some time that an entrepreneurial class emerged to take over some part of the government's role in economic activity.

In some East Asian countries the government nurtured and assisted the development of this class, and sometimes (as in Japan and Korea) provided it with protection from competition until it achieved international competitiveness.

In some developing countries the middle class which emerged had close ties with the government. This facilitated 'rent seeking' and monopolies through collusion between the middle class and government officials. The system which emerged in such cases did not provide equal opportunities for all to "do business."

Governments' determination of the economic direction of Asian countries was a critical factor in shaping the IR systems which emerged. A direct consequence of this was the emergence of the government as the largest employer - particularly in countries which had some socialist orientation auch as India and Sri Lanka. The government as employer, like any other employer would wish to do, influenced the type of IR insitutions which emerged. The shape of IR was further refined by the particular industrialization and economic strategies adopted in each country, as we shall see.

Socialist and import substitution strategies produced rather different models to ones which emerged in the business-friendly, outward-looking, export-oriented countries. But whatever the economic and political orientations of the governments of Asian developing countries, the common element was that economic development and its imperatives were government, and not entrepreneurial, driven. As aptly remarked, the State's role was not "restricted to 'entrepreneurial assistance' only, but extended to 'entrepreneurial substitution'."

A further difference between Western industrialization and the developing country model is that industrialization in developing countries is still occurring in an environment of a dual economic structure, where there is both a large rural/agricultural base and a large and frowing informal sector. Even today in many rapdily industrializing countries such as Thailand and Indonesia the rural and informal sectors continue to be substantial. Unlike in the West. therefore, a proletariat in the sense of a distinctive working class does not numerically dominate the working population.

This has had an effect on the potential membership of unions and their capacity to bargain with employers on equal terms. In some cases the balance has been restored by the political activities of, and political support given to, unions by some former socialist-oriented governments. In other cases the government encouraged (some may say created) a union which the government could control. This is reflected in the right retained by many Asian governments to intervene in collective bargaining, in the restrictions imposed on bargainable issues (Malaysia and Singapore), and in the right to intervene in disputes through the arbitration process.

The many atypical forms of employment in developing countries do not have their counterparts in the employment relationships covered by Western IR systems, and they cannot easily be accommodated in Western labour law concepts. The relatively low level of literacy and education in some developing countries made it difficult for workers to participate in a Western type system, and this was one (though not the only) reason why trade unions were sometimes led by an educated 'elite'. Workers could not at the early stages of industrialization be expected to produce leaders from within their ranks and function as if they had been used to a pluralist, power sharing system.

The approaches to IR in developing countries can be broadly classified into three groups. One goup of countries has had nationalist-populist governments the IR policy of which was

"to encourage the development of the labour movement in a corporatist framework to serve the broad political and economic interests of the state. The central labour organization in such countries is supported by the government ... the government in return for their support increases the job protection and other facilities of the workers.

Most of these steps in favour of the workers are taken without recourse to collective bargaining. To ensure the contnuous support of the trade unions in such systems, the government applied strict control in selecting the leadership and in the structure and functions of the trade unions ... Thus in such a system trade unions are incorporated into the State's administrative structure in a corporatist way"

While this description broadly fits South Asia, in two respects it does not. While India and Sri Lanka have shared a pluralist outlook, there has been, in a legal sense, unrestricted freedom of association, but with influence on union leadership by the political party to which a union may be affiliated. Union supported by opposition political parties have been allowed to function, and in Sri Lanka, for instance, the legal system allows them to canvass freedom of association in the Supreme Court as a fundamental rights issue. In both countries trade unions were involved in the independence movements and, after independence, continuedto enjoy considerable political power and patronage, and sometimes representation in Parliament.

The second group consists of the more authoritarian governments which ensured that unions did not indulge in political activities. In some countries unions were purged of communist elements. The communist movements and insurgencies in some of the South-East and East Asian countries, and the role of unions in this regard, had a decided impact on how unions came to be viewed in these countries. In one way or another unions were controlled so as to prevent them from emerging as a force in opposition to the government. Thus Singapore, Indonesia, Malaysia and the Republic of Korea have favoured recognition of one union.

The third category consisted of the centrally planned socialist countries in which there were no employers as the State was the only employer. The union did not play the role of protecting or furthering the interests of employees as workers were not expected to have interests different to those of the government. The unions administered welfare functions and were in effect instruments of State policy.

The conclusion, therefore, is that "there is no single logic of industrialization leading to one particular type of industrial relations system." Consequently IR in Asia should be understood in the historical context of each country or sub-region. IR and human resource policies have played an important role in the economically successful Asian countries viz. the East Asian and some of the South-East Asian countries.

If IR and human resource policies have failed to promote economic development in South Asian countries, it is arguable that the fundamental problem lay in the economic direction that sub-region followed which did not adapt to changing circumstances, and the IR system in a sense mirrored that direction which failed to deliver growth. Unions in some South Asian countries have been in the forefront of the demands for nationalization of industries and enterprises. Acceding to their demands led to a substantial loss of investments and opportunities to modernize national economies.

Influence of Industrialization Policies

The different economic directions taken by Asia are reflected in the post-colonial industrialization strategies as represented by the outward-looking, export-oriented ones of East and South-East Asia and the inward-looking, import-substitution strategies of South Asia. These strategies have had a marked influence on IR and human resource policies of the different sub-regions. The import-substitution protectionist countries tended to have a labour protection policy based on a highly legalistic IR system in which labour costs were not a very significant consideration, so that productivity, skills development and other conditions necessary to face competition in export markets received little attention.

Countries which concentrated on equity through labour protection were characterized by relative inefficiency, inflexibility and a failure to recognise in time the changes needed to successfully compete in a globalized environment. South Asian IR is characterized by union pluralism, politicization and multiplicity of unions, and in some cases, by extreme labour protection and inter-union rivalry as in India and Sri Lanka, all of which prevented the development of cooperative bipartite relations or meaningful tripartism. IR issues have been heavily influence by political considerations, preventine the development of stable IR.

In a sense, the relative political instability of the region has been reflected in IR as well. However, some of these highly regulated IR systems are gradually adapting to more flexible approaches, and there are indications of a greater desire than hitherto to establish tripartite and bipartite dialogue and cooperation. India, with arguably the most inflexible IR system in Asia, is experiencing pressures to move from a protectionist to a more facilitative, deregulated and flexible system, brought about by economic changes and new industrialization policies.

On the hand labour costs and foreign investment were important factors in the policies of South-East and East Asian countries which depended on the manufacture and export of low cost goods. In some of these countries the industrialization strategy influenced the attitude towards unions. Different strategies were adopted in different countries (Malaysia, Indonesia, Republic of Korea, Japan, Singapore) to remove Communist or left-oriented influence from trade unions. By contrast Communist or left-oriented unions have had considerable influence in Sri Lanka and India ever since those countries gained independence, partly due to the pluralistic political system which they joined and partly due (initially) to their close association with the movement for independence.

With the rapid economic development of several South-East and East Asian countries, greater reliance is now placed on technology, skills, a flexible workforce, employee involvement and cooperation (all of which of course Japan had achieved quite some time ago) necessary for competitve advantage based on productivity and quality. Labour costs, though still important in South-East and East Asian countries are being overshadowed by these other considerations. Indeed, IR systems which hitherto provided little scope for unions to function as they do in market economies, may find it necessary to relax control over union activities in order to facilitate changes in enterprises with the least possible friction.

Malaysia is one of several examples of countries whose industrialization policies affected the IR system. In the early 1980s Malaysia shifted from an import substitution to an export-oriented policy which depended on attracting foreign investment. This shift in economic policy was accompanied by several State interventions in IR intended to attract foreign enterprises and enhance the competitiveness of enterprises. Union structures were changed to encourage enterprise unions, mergers of several national federations of unions were barred, unions were not allowed in the electronics industry, and in order to reduce costs, changes were made in the definition of wages and overtime. Strikes in essential services were prohibitted and the requirement of registration of unions was used to prevent the proliferation of unions.

In the 1990s due to a tight labour market and increased labour costs resulting in investors seeking other countries for their low wage cost activities, Malaysia is concentrating on skills development to attract more technology-based investment. The prohibition of unions in the electronics sector has been removed. Thus the State's role was the dominant one in determining the direction of economic development, and of the IR system thought to be necessary to underpin economic policies.

The Role of Asian Governments

In several South-East and East Asian countries the State - in some cases to the exclusion of unions - regarded itself as the protector of workers and their welfare. The State' splans for growth did not include the practice of pluralism which was seen as being inimical to economic development. Rather than banning unions, they were made adjuncts to the government to ensure the non-emergence of independent unions which could challenge the government's authority or economic plans. Some such governments paid only lip service to the freedom of association and the right to bargain collectively. The State's control of IR has been the strongest in countries in which one party has succeeded in remaining in power for a long period of time, where there have been military regimes, where the government has assumed the primary role in economic development and has exercised a certain degree of authoritarianism, or in socialist countries now moving towards a market economy.

State intervention in the labour market has been pervasive in several East amd South-East Asian countries where the State has viewed itself as the chief architect of economic development and stability. Sei-Park points out that at its inception the Republic of Korea was favoured with an educated workforce but lacked capital. The State's strategy at the time promoted labour-intensive industries with low capital needs, but with the emergence of an entrepreneurial class with capital to invest, the emphasis shifted to heavy industries and investment in developing the necessary skills. The main objectives of Korea's IR policy were industrial peace reflected in the absence of work stoppages, etc and the creation of a favourable labour market climate to achieve economic growth.

Consequently in Korea, as in some other countries in the region which have achieved rapid economic growth, interest groups such as trade unions were not permitted to stall through labour agitation the implementation of policies the government had mapped out to achieve such growth. With rapid economic progress and a well educated workforce the democratic reforms implemented in Korea in 1987 had become inevitable, and the relaxation of decades of union control were followed by a period of militant labour disputes. According to Park the choice facing Korea is between liberal pluralism and liberal corporatism. He believes the second choice is the more likely one, given the relative inexperience of unions in collective bargaining and the Korean tradition of according to the State the main role in regulating society.

Instead of seeking to control the other actors in the IR system, the State in Japan, as pointed out by Kazuo Sugeno, promoted labour-management dialogue, cooperation and stability through procedures for labour dispute adjudication emphasizing employment security, bargaining and joint consultation mechanisms, and by promoting the consolidation of four major trade union confederations into one body. At the national level it has effectively involved employers and workers in consultations in the formulation of labour policies through their participation in trilateral councils. But even Japan in the 1950s witnessed a period of extreme union militancy, which resulted in measures to purge the union movement of its left-oriented elements. It is necessary to draw a distinction between Japan and Korea on the one hand and South East Asia (ASEAN) on the other in regard to their development model. The Japanese and Korean models were based on nurturing and developing national 'champions', with their economies being relatively less open to foreing investment. Nor did they allow financial markets to develop and foreing investment in them. Entrepreneurs had to seek credit from banks whose policies were dictated or influence by governments. The ASEAN countries (and Hong Kong) were more open (in the case of Singapore and Hong Kong completely open) to foreing direct investment.

They were also more inclined to allow financial markets to develop and to allow foreing investment in them. The attempts by South-East Asian countries to emulate the Korean and Japanese examples of developing national 'champions' (e.g. automobiles in the case of Malaysia and aerospace in the case of Indonesia) have been at best, minimal. China and India appear to be taking the path followed by South-East Asia rather than by Korea and Japan. This choice was probably influenced partly by the fact that they are seeking to change their development strategies at a time when investment barriers are disappearing, international capital markets are flourishing, and a free market economy in which governments intervene less has become more commonplace.

While the role of governments in Asian countries has shaped the IR system, in some the role of trade unions has also been an important determinant of IR outcomes. In India and Sri Lanka, for instance, trade unions and many of their leaders were in the forefront of, or identified with, the movements for national independence. Union leaders were either politicians or they enjoyed political patronage. This had a profound influence on post-independence governments which, because they were democratic, sought the support of the working population through trade unions.

The latter were sometimes led by well-educated middle class elite who were able to obtain in return concessions in the form of obligations on employers through legal prescriptions. Representation of trade unions in Parliament enabled unions to push for labour protection legislation. Trade unions, lacking industrial strength, derived distinct advantages from political leadership of unions but, in many cases, failed to produce leaders from within their own ranks. While in several Asian countries only one union has been 'encouraged' (e.g. Singapore, Indonesia), in others (Sri Lanka and India) union pluralism has been recognized from an early stage.

In Asia, therefore, it is not only industrialization policy per se but also the overall role of the government in economic development and the political complexion of the government (whether democratic or authoritarian) which have shaped IR. The core of IR in industrialized market economies is the ability of managements and unions to negotiate terms and conditions of employment relatively free of State control or intervention. In many developing countries this freedom has been substantially less, with the core of the system being the State's power or influence, exercised either through legal control or administrative action. This influence of government in IR has, in several Asian countries, been also prompted by the State's involvement in business and the fact that the State has been the largest - or a large - employer.

The State also assumed the responsibility for ensuring that the two sides of industry act in a manner consistent with the objective of accelerated economic development. Where there is no agreement between the two social partners - and an IR system must pre-suppose the existence of disagreement - a third party has to prescribe the rules which voluntary action has not been able to bring about.

Many Asian governments have, either directly through the exercise of their political or administrative power, or indirectly through Stated-created institutions such as courts and tribunals, prescribed the interests tom be protected, how they are tom be protected and the limits within which the parties may act. Interventions may be the result of demands made by labour or management, each seeking protection by the State of its interests. In South Asia it is the interests of labour which, by and large, have been protected with little consideration given to business interests. The South-East and East Asian business interests have not been sacrificed for the purpose of protecting labour (the latter may admit that their standards of living have substantially improved but claimed that their freedom of action has been limited).

Therefore the legal and administrative interventions of governments have, in a sense, been public policy statements which in South-East and East Asia underlined accelerated economic development, and in South Asia contributed to economic stagnation. The whole of Asia (other than perhaps Japan) has experienced a high degree of State intervention in and control of industrial relations, but the objectives have differed principally between South Asia and the other sub-regions.

Restrictions exist in several countries to ensure that management prerogatives are not eroded, or that industrialization and investment strategies are not impeded. In Singapore and Malaysia, transfer, promotion, retrenchment and lay-off and work assignments are considered management prerogatives and are excluded by law from the scope of collective bargaining. No bargaining on the introduction of new technology is possible in Taiwan. In Singapore and Malaysia collective agreements require certification by the Industrial Court, which is entitled to refuse certification if the provisions are harmful to the national interest. Government intervention is sanctioned in Korea, Japan, Thailand and the Philippines when industrial disputes endanger the national economy. Concessions to foreign investors have been afforded in special economic zones. In some cases limits have been placed on negotiations on terms and conditions of employment in certain types of industries - in 'pioneer' industries in Malaysia and in 'new industrial undertakings' in Singapore.

Strikes have been restricted through several measures. They have been prohibited in essential industries which term has sometimes been defined broadly. A strike can be pre-empted, delayed or rendered illegal in several ways - in Malaysia through the conciliation and arbitration process, or a cooling off period in Korea. Interestingly, in Sri Lanka, a strike can be rendered illegal by a reference of the dispute to an industrial court or to an arbitrator, but this has only very rarely been done even when strikes have been by unions opposed to the government.

In several Asian countries restrictions exist on strikes in the public sector, and strikes are sometimes prohibited if they are a 'secondary' form of trade union action. The several institutional arrangements which exist to channel individual grievances (such as labour courts) and to settle collective disputes (such as conciliation, industrial courts, arbitration) have tended to reduce the need for employers to resort to collective bargaining for which there is, in any event, less scope in an environment of low union rates. By contrast in some South Asian countries such as India and Sri Lanka collective bargaining can take place, and strikes can commence, on practically everything which is connected with the terms and conditions of employment.

Economies in Transition

In Asian economies in transition (China, Mongolia, Viet Nam, Laos and Cambodia) the governments are seeking to establish a labour law system relevant to a market economy. Viet Nam already has a Labour Code, and China is in the process of enacting several laws including one covering collective bargaining contracts. There were hitherto no IR in these countries as known in a market economy, as there were no private employers (or employers' organizations), and employees were not expected to have interests different from those of the employer (the State) as they were considered to be the owners of the enterprises. Decisions were made not so much by managers as by the State.

These countries are now seeking appropriate IR 'models'. Employers in these economies will need to develop the expertise necessary to persuade the other two contituents that the labour law framework should not be too regulated so as to deprive enterprises of the flexibility which will be needed to adapt to changes when these economies have to move to the next stage of economic development. Much will depend on whether the governments, which are interventionist in most aspects, prescribe minimum rules, or whether they will seek to regulate most aspects of the employment relationship. In Viet Nam the law makes it obligatory for a union to be established in each enterprise, so that voluntary unionism is hardly in evidence there.

In Mongolia a law of the early 1990s requires employers to enter into collective bargaining contracts, and not merely into collective bargaining. In some of these countries employers' organizations can be established only if favoured by the governments. However, the governments in these countries are in favour of having an employers' organization, though their role is not viewed in the same way as in a market economy. Employers remain the weaker of the three constituents,and currently also lack the IR expertise to effectively influence legislative outcomes.

Some Evidence of Convergence of Industrial Relations

For reasons we have noted, IR in several Asian countries have been a hybrid - using many Western concepts such as trade unionism, freedom of association, collective bargaining and tripartism, but adapting them to suit particular industrialization/economic policies as well as political/social philosophies of governments. Sometimes even legislative models introduced during British colonial times in different sub-regions have served different purposes. A case in point is the conceptof trade union registration, which helped to support almost unrestricted freedom of association and union multiplicity in India and Sri Lanka, but was used elsewhere (as in Malaysia) to promote a single union structure, or at least to prevent unrestrained multiplicity.

Four circumstances suggest that while IR in Western market economies are undergoing change at least of emphasis, many Asian countries are looking at several aspects of at least Western-influenced IR practices. The first - and general one - is that globalization entails not only internationalization of production and markets, but together with information technology it is resulting in some degree of globalization of ideas and values. This has been given a further impetus in recent times by the debate in relation to the World Trade Organization on whether trade and labour rights should be linked, and the Asian view that labour rights are a matter to be considered within the International Labour Organization.

In addition, the emergence of a better educated and increasingly influential middle class - a process which is nowhere near completion - more open to an outward orientation, will probably contribute to a greater convergence of ideas, including ideas on fundamental rights. Itis not unlikely that Asians will, in due course, come to accept the premise that economic integration, to be truly effective, requires a degree of social cohesion through some shared values among those participating in the global economy. Higher levels of education among employees will probably add to this process. On the other hand, a different view has been advanced on this aspect of the matter which merits consideration:

"...Our preliminary observations do not lend much support to the hypothesis that democratization necessarily fuels the transportation of IR/HR arrangements. It appears that more important explanatory factora are: the dynamics of product (and labour) markets; action by large (and in many cases international) trend-setting companies harnessing the socio-cultural resources available to meet the dynamics of these markets; action by governments (not necessarily ones that have become significantly more democratic."

Second, Asian employers are being increasingly influenced by Western management concepts and techniques, though at the same time seeking to preserve some of their own values. The need for change in the management of enterprises, especially to adapt to the global environment, is increasingly recognized by Asian business leaders.Changes in management practices, as in the case of family businesses which may need to introduce more professional management to operate on a more global scale, could significantly influence employment relations - particularly through the medium of human resource management.

A third circumstance is the pre-occupation of Asian economies in transition to create a suitable IR and labour law system. The interest in using Western market economy concepts adapted to national conditions is evident from, for example, the information and assistance these countries are seeking from the ILO. The following are examples:

- Requests to the ILO for programmes on tripartism
- Legislative activity in China, for example, has been accompanied by requests for information (through workshops and study tours) about IR in market economies outside Asia as well.
- Requests for training in negotiation/collective bargaining (a concept previously non-existent in such countries) and negotiation skills.
- Requests relating to dispute prevention and settlement mechanisms which have basically originated from IR systems in Western market economies. This include workplace and national level mechanisms.

Fourth, while trade union membership in several industrialized market economies is declining, in the Asian industrializing countries trade unions, if not exactly increasing their membership, are reflecting through their activities a degree of labour unrest. Recent events in the Republic of Korea and Indonesia are illustrative, and indicate a move towards greater union pluralism. In economies in transition the pre-existing union continues to function as part of the government apparatus. It is too early to say whether it will ultimately function as an autonomous group. On the whole it would seem that employees wish to have a greater say in regard to their terms and conditions of employment, often through their representatives.

In sum, there seems to be scope for some form of convergence of IR systems, even if not through a transplant of Western systems which are themselves undergoing change. A substantial influence on employment relations may be exercised by HRM, made possible also by the relatively low unionization rates in Asian countries, though the existence of unions should not be a barrier to the introduction of HRM policies and practices. Labour unrest is most likely to surface in countries in which economic growth has been accompanied by widening income disparities, and in countries with highly politicized trade unions and poor attention to mechanisms which promote better workplace cooperation.

Human Resource Management As A Strategy

As we have seen, the model of an enterprise during the greater part of this century was dominated by the concept of 'strategy, structure and systems', which had implications for relations at the workplace. We have also noted that this is changing.

A group of people become an organization when they cooperate with each other to achieve common goals. Communication among them is therefore important. But people have individual motivations which often differ from the corporate goals. An effective organization is one which succeeds in getting people to accept that cooperating to achieve organizational goals also helps them to achieve their own goals provided they are adequately rewarded through extrinsic and intrinsic rewards. This is achieved primarily through leadership and motivation.

Employers therefore increasingly view human resource management from a strategic perspective, and as an appropriate means through which the chasm between organizational and individual goals can be narrowed. As it has been aptly observed:

Part of the problem is that we have split off human resource management from the general management problem, as if there were some other kind of management other than human resource management. As long as organizations are based upon the coordinated action of two or more people, management is by definition human resource management.

Despite the proliferation of writings and studies on HRM, there is a wide gap between the rhetoric and the reality, though the gap has been narrowing in the 1990s. There is as yet inadequate research to ascertain the extent to which practice matches corporate policy statements, and the impact of HRM policies and practices on employee behaviour and morale. To have a major impact on enterprises, HRM has to be diffused across an economy, rather than remain islands of excellence. Nevertheless, promoting excellent models of HRM stimulates interest in better people management.

From Personnel Management To Human Resource Management

The increasingly important role of HRM is reflected in the transformation of the personnel management function from one of concentrating on employee welfare to one of managing people in a way which matches organizational and individual goals and providing employees with intrinsic and extrinsic rewards. Therefore today

"far from being marginalised, the human resource management function becomes recognized as a central business concern; its performance and delivery are integrated into line management; the aims shift from merely securing compliance to the more ambitious one of winning commitment. The emplouee resource, therefore, becomes worth investing in, and training and development thus assume a higher profile. These initiatives are associated with, and maybe are even predicated upon, a tendency to shift from a collective orientation to the management of the workforce to an individualistic one..Accordingly management looks for 'flexibility' and seeks to reward differential performance in a differential way. Communication of managerial objectives and aspirations takes on a whole new importance."

What separates or distinguishes HRM from the traditional personnel function is the integration of HRM into strategic management and the pre-occupation of HRM with utilizing the human resource to achieve strategic management objectives. HRM "seeks to eliminate the mediation role and adopts a generally unitarist perspective. It emphasizes strategy and planning rather than problem solving and mediation, so that employee cooperation is delivered by programmes of corporate culture, remuneration packaging, team building and management development for core employees, while peripheral employees are kept at arm's length."

HRM strategies may be influenced by the decisions taken on strategy (the nature of the business currently and in the future) and by the structure of the enterprise (the manner in which the enterprise is structured or organized to meet is objectives). In an enterprise with effective HRM polices and practices, the decisions on HRM are also strategic decisions influenced by strategy and structure, and by external factors such as trade unions, the labour market situation and the legal system. In reality most firms do not have such a well thought out sequential HRM model. But we are considering here is also effective HRM, and thus a model where HRM decisions are as strategic as the decisions on the type of business and structure.

At a conceptual level the interpretations of HRM indicate different emphases which lead to concentration on different contents of the discipline. The various distinctions or interpretations indicate that HRM "can be used in a restricted sense so reserving it as a label only for that approach to labour management which treats labour as a valued asset rather than a variable cost and which accordingly counsels investment in the labour resource through training and development and through measures designed to attract and retain a committed workforce.

Alternatively it is sometimes used in an extended way so as to refer to a whole array of recent managerial initiatives including measures to increase the flexible utilization of the labour resource and other measures which are largely directed at the individual employee. But another distinction can also be drawn. This directs attention to the 'hard' and 'soft' versions of HRM. The 'hard' one emphasizes the quantitative, calculative and business-strategic aspects of managing the headcounts resource in as 'rational' a way as for any other economic factor. By contrast, the 'soft' version traces its roots to the human-relations school; it emphasizes communication, motivation, and leadership."

There are several ways in which HRM has changed earlier attitudes and assumptions of personnel management about managing people. The new model of HRM includes many elements vital to the basic management goal of achieving and maintaining competitiveness.

First, HRM earlier reacted piece-meal to problems as they arose. Effective HRM seeks to linke HRM issues to the overall strategy of the organization, with the most effective HRM policies and practices integrated into such corporate policies and strategies to reinforce or change an organization's culture. Integration is needed in two senses - integrating HRM issues in an organization's strategic plans and securing the acceptance and inclusion of a HRM view in the decisions of line managers. The HRM policies in respect of the various functions (e.g. recruitment, training, etc.) should be internally consistent. They must also be consistent with the business strategies and should reflect the organization's core values. The problem of integrating HRM view business strategy arises, for example, in a diversified enterprise with different products and markets. In such cases it is difficult to match HRM policies with strategies which could vary among different business activities, each of which may call for different HRM policies.

Second, building strong cultrues is a way of promoting particular organizational goals, in that "a 'strong culture' is aimed at unting employees through a shared set of managerially sanctioned values ('quality', 'service', 'innovation' etc.) that assume an indentification of employee and employer interests." However, there can be tension between a strong organizational culture and the need to adapt to changed circumstances and to be flexible, particularly in the highly competitive and rapidly changing environment in which employers have to operate today. Rapid change demanded by the market is sometimes difficult in an organization with a strong culture.

IBM has been cited as a case in point. Its firmly-held beliefs about products and services made it difficult for it to effect changes in time, i.e. when the market required a radical change in product and service (from mainframe, customized systems, salesmen as management consultants to customer-as-end-user, seeking quality of product and service) to personal computers (standardized product, cost competition, dealer as customer).Nevertheless, in the long term a strong organizational culture is preferable to a weak one:

"Hence it could be said that the relationship between 'strong' cultures, employee commitment, and adaptability contains a series of paradoxes. Strong cultures allow for a rapid response to familiar conditions, but inhibit immediate flexibility in response to the unfamiliar, because of the commitment generated to a (now) inappropriate ideology. 'Weak' cultures, in contrast, when equated with ambiguous ideaologies, allow flexibility in response to the unfamiliar, but cannot generate commitment to action. Yet strong cultures, through disconfirmation and eventual ideological shift may prove ultimately more adaptive to change, assuming the emergence of a new strong yet apprpriate culture. This may be at the cost of a transitional period when ability to generate commitment to any course of action - new or old - is minimal."

Third, the attitude that people are a variable cost is, in effective HRM, replaced by the view that people are a resource and that as social capital can be developed and can contribute to competitive advantage. Increasingly, it is accepted that competitive advantage is gained through well-educated and trained, motivated and committed employees at all levels. This recognition is now almost universal, and accounts for the plausible argument that training and development are, or will be, the central pillar of HRM. Thus

"The existence of policies and practices designed to realise the latent potential of the workforce at all levels becomes the litmus test of an organization's orientation."

Fourth, the view that the interests of employees and management or shareholders are divergent and conflictual - though substantially true in the past - is giving way to the view that this need not necessarily be so. As organization which practices effective HRM seeks to identify and promote a commonality of interests.

Significant examples are training which enhances employment security and higher earning capacity for employees while at the same time increasing the employee's value to the enterprise's goals of better productivity and performance; pay systems which increase earnings without significant labour cost increases, and which at the same time promote higher performance levels; goal-setting through two-way communication which establishes unified goals and objectives and which provides intrinsic rewards to the employee through a participatory process.

Fifth, top-down communication coupled with controlled information flow to keep power within the control of management giving way to a sharing of information and knowledge. This change facilitates the creation of trust and commitment and makes knowledge more productive. Control from the top is in effective HRM being replaced by increasing employee participation and policies which foster commitment and flexibility which help organizations to change when necessary. The ways in which the larger Japanese enterprises have installed participatory schemes and introduced information-sharing and two-way communication systems are instructive in this regard.

In enterprises which tend to have corporate philosophies or missions, and where there are underlying values which shape their corporate culture, HRM becomes a part of the strategy to achieve their objectives. In some types of enterprises such as ones in which continuous technological change takes place, the goal of successfully managing change at short intervals often requires employee cooperation through emphasis on communication and involvement. As this type of unit grows,

"If there is strategic thinking in human resource management these units are likely to wish to develop employee-relations policies based on high individualism paying above market rates to recruit and retain the best labour, careful selection and recruitment systems to ensure high quality and skill potential, emphasis on internal training schemes to develop potential for further growth, payment system designed to reward individual performance and cooperation, performance and appraisal reviews, and strong emphasis on team work and communication ... In short, technical and capital investment is matched by human resource investments, at times reaching near the ideals of human resource management."

If, as is often the case, (the UK is a good example(, enterprises are dominated by financial issues, HRM will not be a part of the central strategy of such enterprises.

Human Resource Management In Achieving Management Objectives

HRM has three basic goals which contribute to achieving management objectives. The first is integration of HRM in two senses: integrating HRM into an organization's corporate strategy, and ensuring an HRM view in the decisions and actions of line managers. Integration in the first sense involves selecting the HRM options consistent with (and which promote) the particular corporate strategy.

The option is determined by the type of employee behaviour expected (e.g. innovation) needed to further the corporated strategy. For instance, the HRM policies in relation to recruitment, appraisal, compensation, training, etc. differ according to whether the business strategy is one of innovation, quality enhancement or cost reduction.

A strategy of innovation may require a pay system less influenced by market rates but which rewards creativity, and the pay rates would even be low so long as there are ways of making up the earnings package. A cost reduction strategy may lead to pay rates being strongly influenced by market levels. Similarly, training and development would receive less emphasis in a cost reduction strategy than in one where the objective is innovation or quality. But such integration is difficult without securing the inclusion of a HRM view in the decisions and practices of line managers. This requires that HRM should not be a centralized function.

A second goal of HRM is securing commitment through building strong cultures. This involves promoting organizational goals by uniting employees through a shared set of values (quality, service, innovation, etc.) based on a convergence of employee and enterprise interests, which the larger Japanese enterprises have been particularly adept at. A third goal of HRM is to achieve flexibility and adaptability to manage change and innovation in response to rapid changes consequent upon globalization. Relevant to HRM policies in this regard are training and multi-skilling, re-organization of work and removal of narrow job classifications. Appropriate HRM policies are designed, for instance, to recruit, develop and retain quality staff, to formulate and implement agreed performance goals and measures, and to build a unified organizational culture.

The Increasing Interest in Human Resource Management

The 1990s in particular have witnessed an increasing interest in HRM, including in Asia. This is likely to transform rhetoric into more widespread practice and implementation of policies. Among the factors which have contributed to the increased interest in HRM are the following:

- The shift from addressing issues relating to the employment relationship at levels external to the enterprise to addressing them at the enterprise level. This shift is evident even in countries which have operated centralized IR systems to determine basic terms and coditions relating to the employment relationship (e.g. Australia, New Zealand, Sweden). This shift is associated with the idea propogated by employers that in an increasingly globalized environment competitiveness is won or lost at the level of the enterprise and industry. As such, changes that are needed to make enterprises competitive can best be effected by matching organizational and individual goals, which requires action at the enterprise level. Since by its very nature, HRM operates at the enterprise level, it is being viewed by employers as the preferred method.
- Since much of IR activity in the past has focused on relations outside the workplace, IR is seen as less relevant to management objectives.
- The formulation and implementation of HRM policies are not necessarily dependent on the existence of a union, unlike traditional IR which is based on the premise of the existence of unions. For instance, collective bargaining usually assumes a union as the bargaining agent of employees. Many of the areas of HRM (selection and recruitment, leadership and motivation, employee development and training, employee retention) are implementable even where there is no union. However, this is not to suggest that unions should not be involved where they exist, and as we shall see, they have worked best in a unionized setting.
- The increasing examples of excellence in HRM have generated an interest in it as a means of achieving management objectives.
- The traditional role of personnel managers has failed to exploit the potential benefits of effective management of people. Nor did personnel management form a central part of management activity.
- Decline in unionism has opened the way for managements to focus more on individuals, rather than on collective issues.
- Many important aspects of HRM (e.g. commitment, motivation, leadership) emanate from the area of organizational behaviour, and

underline management strategy. This has provided the opportunity to link HRM with organizational behaviour and management strategy.

The Conflict Between Industrial Relations and Human Resource Management

The two basic issues are how does HRM challenge IR, and how can the conflict, if any, be resolved so that the two complement each other. This section addresses the first of these two issues. In addressing the first question, it is necessary to identify some of the key differences between HRM and IR(. IR is essentially collectivist and pluralist in outlook, dealing with relations between employers and unions, and between them and the State, which predicates that the outcomes are standardized rules and procedures.

The collective aspect of IR is reflected in some of the central features of an IR system such as freedom of association, collective bargainimg, right to strike, trade unionism, dispute settlement and worker participation in management through union participation.

HRM, on the other hand, does not encompass a third party (the State). It is bipartite but essentially individual focused as is evident from the key HRM subjects such as selection and recruitment, induction, appraisal, development and training, leadership and motivation, and retention of staff through intrinsic and extrinsic rewards.The point is well made by one writer:

“The empirical evidence also indicates that the driving force behind the introduction of HRM appears to have little to do with industrial relations; rather it is the pursuit of competitive advantage in the market place through provision high quality goods and services, through competitive pricing linked to high productivity and through the capacity swiftly to innovate and manage change in response to changes in the market place or to breakthroughs in research and development .... Its underlying values, reflected in HRM policies and practices, would appear to be essentially unitarist and individualistic in contrast to the more pluralist and collective values of traditional industrial relations."

A second point of difference is that IR consists of a large component of rules set by the State through laws, by the parties through negotiated agreements, or by courts or tribunals. HRM deals less with rules than with policies and practices "designed to maximize organizational integration, employee commitment, flexibility and quality of work. Within this model, collective industrial relations have, at best, only a minor role."

A third point of differernce, flowing from the earlier mentioned differences, is that the pluralist outlook of IR assumes (of course correctly) a potential for conflict between the two parties, or between one of them and the State, flowing from difference interests. IR seeks to balance these interests through means directed at the 'collective' e.g. collective bargaining or other similar procedures as well as dispute settlement procedures. A concession is made to the individual in the form of grievance handling procedures. HRM is unitarist and sees - or at least tries to achieve - a commonality of interests. It views problems as emanating from reasons internal to the enterprise (e.g. poor people management), requiring them to be addressed through internal, not externally imposed, policies. HRM is the management of human resources rather than collective relations, and is therefore enterprise focused.

From these differences flows a fourth difference. HRM involves the individualization of the employment relationship, whereas participation in IR involves unions rather than individuals directly. In terms of securing commitment HRM addresses the issue both individually and collectively. In the case of pay and rewards, IR has generally emphasized the collective, and therefore standardization, as leading to equity. HRM sees equity and efficiency flowing from differentiation based on performance (group or individual_ and skills application; hance the development of performance and skill based pay systems.

This difference is further heightened by the flexibility (working hours, types of contracts, functional,pay, etc.) claimed by employers to be needed to enable them to adapt rapidly to change. This claim therefore breaks down the standardization inherent in traditional IR approaches to wages, contracts, functions, working hours. It also creates tensions within enterprises with unions, which have preferred the employment relationship to be governed by standardized terms and conditions of employment. For example, in IR pay is based largely on job evaluation and cost of living, while HRM seeks to introduce a performance element into pay.

A fifth difference between IR and HRM is the more integrated approach of HRM which places it closer to corporate planning and strategy than IR, which has always been only at the periphery of corporate planning. This integration also involves greater involvement of line managers in HRM, concentrating as it does on the individuals in the line manager's department, whereas IR has been regarded as a specialist's function.

A sixth - and a major difference or more accurately a major potential point of conflict between employers and unions and between HRM and IR - is the employee loyalty and commitment HRM seeks to harness. The issue sometimes resolves itself into whether dual allegiance is possible i.e. commitment to the goals and values of the organization on the one hand, and commitment to the trade union on the other. In principle, there should not be a conflict, as is evident from the situation in the larger Japanese enterprises where HRM and IR have co-existed, unions have been involved in HRM policies (facilitated by enterprise unionism), and consultation and communication systems have considerably contributed to the development of common understandings and goals among employees, unions and managements on the relationship between enterprise growth, competitiveness and their own welfare. Dual loyalty is possible only where unions and their members share some goals in common with management.

Seventh, the emergence of new categories of employees has heightened the distinctions between IR and HRM. Industrialized countries are witnessing the emergence of knowledge workers who, as the professional core who "own the organizational knowledge which distinguishes that organization from its counterparts. Lose them and you lose the organization." It is their ability to "allocate knowledge to productive use" that will determine the competitiveness of organizations in the future. Knowledge workers are costly, often in short supply, owe little allegiance to countries, and are highly mobile because they can move even across national borders with relative ease to location where their knowledge-based skills are most in demands and the rewards the greatest.

Since they prefer an envionment conducive to creativity, innovation and the flowering of their skills and talents, enterprises with human resource systems which approximate as closely as possible to the ideal model of HRM are the ones that attract them. The individualization thrust of HRM is more suited to their aspirations, rather than the rule-bound, standardized and collectivist approach of IR. They do not expect to be treated under a standardized system - under a collectively negotiated pay system for instance - but expect performance to be recognized and rewarded.

The proliferation of other categories of employees - part-time, home workers, contract workers - do not fit the traditional IR models developed in an age of mass production to fit the notion of all employees working within an enterprise, rather than for one. Several other issues arise in connection with IR and HRM such as whether HRM is practised as an anti-inion or union avoidance strategy, the effect of unionization on HRM, and union views on HRM. These are outside the scope of this Paper. In summary, many employers see an incompatibility between IR and HRM for reasons such as the following:

- IR considers the mass rather than the individual
- In IR pay determination has traditionally been on criteria different to objectives sought to be achieved by HRM
- IR seeks to reconcile conflict; HRM to match goals
- In IR communication with employees is through unions; in HRM it is not necessarily so.
- IR has traditionally promoted standardization, whereas HRM is more concerned with flexibility
- Some of the central issues in HRM such as leadership, motivation and training either do not form part of IR (leadership and motivation) or have been a relatively incidental focus (training).

Harmonizing Industrial Relations

In 1993 the European employers' organizations at their annual general meeting addressed the issue of HRM, and the following summary reflects their views:

"The key to competitiveness is quality. And quality depends more on the commitment of individuals than on their acquired technical skills; more on the way these individuals behave and their tema spirit than on the passive execution of orders received. Regulations - be they internal to the enterprise or imposed by the legislator - plague innovation and have a negative effect on motivation. Good human resource management lies in the behaviour of each employee within an enterprise ...

It applies to individual men and women.... The classic approach to industrial relations is entirely different, with its peculiarity being collective bargaining - which, by definition, does not consider the individual but the mass. The status of a worker is defined by a few general criteria such as his professional category and possibly his seniority and his age. Remuneration, for example, is calculated on the basis of a few simple elements, which do not take into consideration personal behaviour and individual performance.

Work provided is considered purely from the quantitative angle. The future of collective bargaining, therefore, depends on the extent to which it can take into account the demands for individualization which modern methods of human resources management imply.. This hsould not of course lead to arbitrariness... Legislation and regultation imposed by the State should also leave a sufficient margin of flexibility to alow for adaptation."

Reconciling or harmonizing the apparent incompatibility between IR and HRM is a major challenge faced by IR. If it is to take place it requires integration which involves expanding the frontiers of both i.e. HRM should take into account the external environment more than it does at present, and IR should focus far more than it does on workplace relations and recognize that it should not consist largely of collective relations. Alternatively, it requires the two to operate as dual or parallel systems, which I sless advantageous, because this is less likely to expand the frontiers of either.

There are some pre-conditions to harmonization:

- Changes in both management and union attitudes
- Acknowledgement of the link between employee development and enterprise growth
- Recognition that employer and employee interests are not only divergent but also common e.g. productivity is an important issue to both
- Both HRM and IR should be prepared to accommodate the other. At present IR seems to view HRM as its nemesis.
- Unions would need to be more willing to involve themselves in HRM, and not over-emphasize their national agenda. Otherwise employers will have free rein in ppromoting HRM at the expense of IR. This also involves training of unions in HRM. Where union structures change, as happened in Japan when enterprise unions became an important features, the potential for such involvement is greater.

Union cooperation would depend to a large extent on whether in a particular enterprise HRM is perceived as a union avoidance strategy. Union involvement can be critical, since research appears to indicate that HRM initiatives have worked most effectively in unionized settings.

- Changes in IR thinking
- redesigning collective bargaining to accommodate more workplace issues (as is already occurring)
- less adversarial relation
- IR needs to open its doors to other social disciplines (e.g. organizational behaviour and psychology, industrial sociology) with consequent attention to team work and new forms of work organization.
- IR would have to recognize that communication in an enterprise need not necessarily be only effected collectively. Dual communication systems (asin the case of joint councils in Japan) are possible and can complement each other.
- IR has to embrace the whole employment relationship, and not only its collective aspects.
- Managements should themselves be willing to involve unions in HRM initiatives. Workplace cooperation mechanisms, for instance, tend to be more effective where, in a unionized environment, unions are consulted.
- A more strategic perspective of IR needs to be developed, going beyond including traditional objectives such as distributive justice. Such a perspective would need to place less emphasis on standarization and to espouse productivity and competitiveness.

The other solution is that HRM will more or less take over IR's role within the enterprise, and IR will continue to operate at a different level outside the workplace, with the State continuing to play a role, albeit a diminished one. The feasibility of this solution depends on a consensus that some common standard rules are necessary in any social system.

Future of Human Resource Management

If HRM is not to remain more in the realm of rhetoric with wide disparaties between theory and practice, several things need totake place. First, HRM needs to be diffused across industries and the economy. For this to occur the following conditions need to be satisfied:

- HRM should be an essential part of management education and training (some would say that it should be the essence).
- From this, two important consequences are likely to follow. HRM is likely to be ingetrated into corporate strategies and line managers' functions and decisions. This would reduce the need for HRM

specialists, except at the policy level where they will have a greater voice. Business strategies are then likely to be built less around low cost and low wages, but around the real sources of competitive advantage such as flexibility, quality and customer service.

- Employment policies which support employment security, without which HRM policies, including training, would have little motivational effect. This does not mean guaranteed employment, but a policy which treats termination as a last, rather than a first, resort.
- Learning from international experiences and diffusing the information can have a transforming effect, as was the case when American manufacturing was transformed through in-depth studies of Japanese manufacturing in the automobile industry.
- Substnatial investment in people and the willingness of employers to view the benefits from a long-term perspective - a difficult task in a system which is driven by short-term investor pressure.
- HRM requires to overcome one of its weaknesses, namely, to recognize that the choices available to managements are governed not only by internal but also by external considerations. "Ironically ...students of HRM often begin with the weakening of external labour market institutions and the liberalization of management in the firm as necessary pre-conditions for the adoption of HRM ... the more HRM is seen to be the preserve of each individual firm acting in isolation, the least likely it is that HRM practices will grow and flourish in the wider economy."

FUTURE ISSUES FOR INDUSTRIAL RELATIONS

Continuing Relevance of Industrial Relations

In a globalized environment with businesses, money and people moving with relative ease across borders, the relentless pursuit of competitive advantage at the expense of all else, the disruption of social relationships and stability, the rapdi outdating of knowledge, skills and technology, with learning being a life-long pursuit, and increasing job insecurity, the only certain factor is change and its rapidity.

Poverty worlwide is nowhere near reduction to minimal levels, and on the contrary, is increasing. Many of the benefits of recent changes have benefitted a few, and in many countries income gaps are widening, rather than narrowing. It has been suggested by eminent writers that the world may well be heading towards over-production of goods, food shortages, and environmental degradation.

Before we dispense with institutions or systems which have contributed to social stability, it is worthwhile assessing their continued relevance. Industrial relations is one such.

IR is no doubt undergoing needed changes, but it is by no means irrelevant. Its major contribution was that it facilitated distributive justice and thereby contributed to social stability. Western Europe is probably the best example of an IR system which was underpinned by its social market principles and, by concerning itself with distributive justice and equity, raised the living standards of the majority, thus providing decades of relative social stability. If, as many employers claim, the labour market in that region is too regulated in the context of the changed environment, this does not implly a total abandonment of the system, but only its reform. In fact when we speak of changes in IR in many countries, it does not always imply a radical change, but rather a change of emphasis. For instance, the idea of negotiation on which collective bargaining is based, continues to be valid even if the trend is towards decentralized bargaining. Nor is there anything in HRM that contradicts the value of negotiation.

HRM undoubtedly poses a challenge to IR as we have seen. But a democracy and pluralism are based on the recognition of different interest groups within a society, each acting as a check and balance in relation to the other to prevent a centralization of power.

A system which provides some external regulation of the behaviour of groups must therefore be necessary.

Since HRM is enterprise-focused, there is a need for a system which can deal with issues which arise in the external labour market. At last for those who have no individual bargaining power - and they constitute the majority of the world's working population - traditional IR institutions such as freedom of association, collective bargaining, minimum employment terms (e. g. age of employment, force labour, safety and health, holidays), social security and dispute settlement mechanisms continue to be relevant. Policies need to be formulated on these matters and applied across society. The fact that some traditional IR features may need to be changed, does not imply that they are irrelevant; the need for a greater enterprise focus does not imply the absence of a national focus as well.

IR institutions continue to have the following relevance:

- Collective bargaining, even if it be at the enterprise level, can still help to reduce inequalities in negotiating power
- Freedom of association provides the foundation for the recognition that employees have rights.
- Industrial peace needs to be ensured both by addressing it at the enterprise level, any by providing in the event of their failure, safeguard mechanisms external to the enterprise such as conciliation, courts or tribunals.
- Processes such as tripartism are needed to ensure that the relevant parties have the opportunity to influence labour policy and legislative outcomes.

- The boundaries of action within which parties may act need to be set.
- Social protection through minimumn standards may often be required, whether they relate to children or women, safety and health or superannuation.

Current and Future Issues

In these circumstances the issues which IR will be called upon to address, in particular in Asia, need to be identified. Employers are now compelled to view IR and HRM from a strategic perspective; in other words, not only from the traditional viewpoint of negotiating terms and conditions of employment and performing a personnel and welfare function. IR and HRM are directly relevant to competitiveness, and how they are managed will impact on enterprise performance, e.g.

its productivity and quality of goods and services, labour costs, quality of the workforce, motivation, prevention of disputes and not only their settlement, and aligning employee aspirations with enterprise objectives.

Minimum Wages

In countries which have a legal minimum wage three employer concerns are evident. The first is that minimum wage levels sometimes tend to be fixed on extraneous condierations (e.g. political), or on inadeuate data needed to define the level of wages. The second concern is that such instances have an adverse effect on competitiveness in the global market and on employment creation where the minimum wage is fixed above a certain level (much of the controversy relates to what that level is). Therefore many employers prefer to see the minimum wage, if there is to be one at all, as a 'safety net' measure to uplift those living below the poverty line.

The third concern relates to increases in minimum wages nopt being matched by productivity gains which would help to offset increased labour costs.

Flexible/Performance Pay

Many employers, and even some governments, wish to review traditional criteria to determine pay levels such as the cost of living and seniority. Pay system which are flexible (i.e based on profitability and productivity) so as to be able to absorb business downturns and also reward performance, are receiving considerable attention. One major problem in this regard is how employees and their organizations can be persuaded to negotiate on pay reform. The objectives of pay reform will not be achieved unless reforms are the result of consensual agreements and are a part of a larger human resource management strategy and change in human resource management systems.

"We now pay workers not for output produced, nor even for labour input provided, but simply for time spend on the job." Traditionally wages and pay have been determined through government regulation, minimum wage determination, negotiation with unions, decisions of arbitration and labour courts and the individual contract of employment.

The factors or criteria which have influenced pay and pay increases have include profit (but generally unrelated to individual or group performance). job evaluation, seniority, cost of living, manpower shortage or surplus, negotiating strength of the parties and skills. Performance measures such as productivity or profit related to the performance of a group has been of less importance in determining pay increases. Though skills have been reflected in pay differentials, pay systems have been seldom geared to the encouragement of skills acquisition and application.

Industrialized countries have built their competitive advantage not around low wages, but around clusters of competitive industries in which high earnings and standards of living have been sustained through improved technology, productivity and quality. Many Asian countries now recognize that high technology, productivity and low earnings cannot be combined and sustained over a long period of time.

Many Asian employers are now seeking to sustain their competitiveness through pay increases which are more related to performance measures as a way of absorbing increased labour costs, while at the same time rewarding and motivating employees.

Japan, Singapore and the Republic of Korea, for example, have succeeded in moving to high value-added and technology-based or service activities partly because they had invested in skills development and accepted the fact that higher earnings (in Singapore partly through a flexible pay system) are an essential strategy for entry into the knowledge-based industries of the future.

Increases in real earnings have been made possible because investment in education and skills contributed to productivity enhancement which, in turn, created the capacity to absorb higher earnings. In the 1980s Singapore made a deliberate shift to a high wage economy in order to encourage entry into high value-added activities. Productivity increases are necessary to sustain higher earnings; at the same time there cannot be a long term productivity growth without an increase in real wages.

Traditional pay systems for non-executive staff have generally been characterized by standardization across and within sectors (e.g. government, particular industries) and within enterprises. So long as employers were competing mainly in domestic markets which were protected from foreign competition - in some cases leading to monopolies - the effects of standardization on considerations such as performance, recruitment and retention of good staff, etc. were less felt. Indeed, standardization, while being equitable from the point of view of employees, benefited employers as well by reducing competition based on labour costs.

With the gradual opening up of economies to world trade and foreign investment, local employers are now compelled to compete with enterprises with sophisticated technology, more productive ways of providing goods and services, and the advantage of being global players.

In many instances these foreign enterprises are able to attract the best local talent on terms and conditions beyond the capacity of many local enterprises to pay. With the acceleration of the process of globalization, accompanied by the movement of former centrally planned economies towards market economies, governments and private enterprises have had to compete in the global market by developing competitive advantages, which are affected by costs and quality.

Productivity increase as the measure of performance at the national and enterprise levels, with quality as an intrinsic part of productivity, is becoming the goal of many developing countries as well - now pushed to the forefront by the forces of globalization and the collapse of economic systems which were alternatives to a market economy. Economies which are seeking to progress from low cost manufacturing to highly-skilled and technology-based production need pay systems which not merely recognize skill differentials (as standard pay systems do), but also provide an incentive to acquire skills and multi-skills facilitated by years of careful and correct investment in education and training.

In the area of IR, collective bargaining outside the enterprise is seen by employers as achieving distributive justice in the sense of equality, with outcomes often being based on the bargaining strngth of the parties. It is increasingly viewed as contributing little to productivity and performance. The outcomes often leave employers with little or no capacity to make further payments on account of performance under a scheme. The movement towards decentralization of collective bargaining has been the result of the need to address efficiency and performance issues at the enterprise level. It is natural that with decentralization employers would seek ways to introduce performance criteria into wage increases.

The increase in atypical forms of employment (e.g. homework) which cannot consistently or always be supervised, has also influenced the search for alternative forms of pay.

Traditionally increased earnings were secured and performance rewarded partly through promotions. With limitations on higher positions in the context of organizations becoming less hierarchical in the future, relating part of pay increases to performance would be a way of rewarding performance, other than through promotions. In these circumstances pay systems are increasingly forming a part of human resource management initiatives to achieve enterprise-level objectives and strategies, with more attention being paid to how they fit into the overall human resource management policies of enterprises.

These developments have several implications for pay systems. Employers (and some governments) see that pay increases need to be more than matched by productivity increases if competitiveness is to be achieved or maintained. The relationship between pay, productivity, skills and inflation were understood quite early in Asia by Japan as well as by the newly industrialised economies. Some other South-East Asian countries have also come to appreciate this relationship.

The pressure in the Asia-Pacific region for the recognition of performance criteria in pay determination has not come only from employers. In Singapore the government initiated the move to flexible pay.

In Malaysia the government drew attention in 1988 to the desirability of introducing a flexible pay system. In Fiji the government has been exploring the feasibility of introducing performance pay into the public service, while encouraging employers to do likelwise.

The Minimum Wage Board in Papua New Guinea has, since the early 1990s, been required to relate minimum wages to performance criteria in place of indexation. In recent years the wage determination system in Australia has encouraged employers and unions to negotiate a part of wage increases in the context of productivity and productivity-related improvements. The fundamental shift of industrial relations to the enterprise level and the individualization of the employment relationship in New Zealand have provided ample scope for performance-related pay.

Therefore it is increasingly recognized that performance and skills criteria need to be injected into pay determination; that it cannot be achieved through centralized or macro-level pay determination; and that changes have to be negotiated at the enterprise level. In these circumstances, a major concern for employers will be to negotiate pay systems which

- Achieve a strategic business objective;
- Are flexible in that their variable component could absorb downturns in business and reduce labour costs;
- Are oriented towards better performance in terms of productivity, quality, profit, etc.;
- Are capable of enhancing workers' earnings through improved performance;
- Are capable of reducing the incidence of redundancies in times of recession or poor enterprise performance through the flexible component of pay;
- Are able to reward good performance without increasing labour costs, and
- Are able to attract competent staff.

The types of schemes which fall within the description of performance pay are varied. Broadly speaking, they consist of schemes designed to share or distribute the financial results of enterprise performance with or to employees. In essence, performance pay is based on paying the worker for his or her value, rather than the value of the job. Such schemes fall into four broad categories:

- Individual-based or based on individual performance, such as incentive schemes and sales commissions
- Profit-sharing which applies to all or most of the employees
- Gain-sharing measured by a pre-determined formula, applicable to all groups of employees. The performance measure may be profit or some other objective such as productivity
- Employees share ownership schemes.

Cross-Cultural Management

Asia is a heterogeneous region, characterized by ethnic, cultural, linguistic and religious diversity. Due to substantial increases in investment in Asia by both Asian and Western investors, many employers and unions are dealing with workers and employers from backgrounds and cultures different to their own. Some of the resulting problems and issues (reflected, for instance, in the proliferation of disputes due to cross-cultural 'mismanagement') fall within cross-cultural management. The problems arise due to differences in IR systems, attitudes to and of unions, work ethics, motivational systems and leadership styles, negotiating techniques, inappropriate communication, consultation and participation procedures and mechanisms, values (the basic beliefs that underpin the way we think, feel and respond), expectations of workers and interpersonal relationships.

These cross-cultural management issues in turn pose the following problems:

- What particular IR and human resource management considerations at the regional, sub-regional and country level affect the development of sound relations at the enterprise level in a cross-cultural environment?
- What would be the most effective programmes for this purpose?
- How can investors in Asia familiarize themselves with the environmental and cultural considerations in the recipient country relevant to their managing people at work?
- How could information be collected, analyzed and disseminated?

Dispute Prevention

Most countries (other than those in transition to a market economy) have long-stadnign dispute settlement procedures at the national level (e.g. conciliation, arbitration, industrial or labour courts). Essential as these are, they operate only when a dispute arises. Equally important are dispute prevention through communication, consultation and negotiation procedures and mechanisms operating at the enterprise level. These are not particularly common in many Asian enterprises. A more positive movement from personnel management to strategic buman resource management is called for.

Industrial Relations/Human Resource Management Training

Not many developing countries in the region have facilities for training in labour law and IR - negotiation, wage determination, dispute prevention and settlement, the several aspects of the contract of employment, and other related subjects such as safety and health. More facilities are probably available in human resource management. Since IR has assumed a particularly important role in the context of globalization, structural adjustment and in the transition to a market economy, employers in each country would need to identify what aspects of IR and HRM should be accorded priority, how training in them could be delivered, and what concrete role is expected from an employers' organization.

Balancing Efficiency With Equity and Labour Market Flexibility

During this century IR and the law in industrialized countries have paid considerable attention to the means through which the unequal bargaining position between employees and management can be rectified. The imbalance in their respective positions has been corrected primarily through the freedom of association and collective bargaining. Thereafter the focus in some countries has been more on the relationship between management and labour and their organization rather than on their relationships with the state.

This has been due to the fact that the State has adopted a less interventionist role than in developing countries, based on the premise that regulation of the labour market should, to a large extent, be left to the employers, workers and their organizations.

However, in some Western European countries, Australia and most Asian countries attention has focused on relationships with the State because of the role governments have played in regulating the labour market (through laws and also through labour courts or tribunals), or in directing economic development and industrialization.

Traditional IR view labour problems as arising due to employers wishing to use resources productively and to generate profit, while employees wish to maximise their return on labour. The State intervenes for a variety of reasons.

The setting in which IR developed was conditioned by the national environment - political, economic, social and legal. But today the conditioning environment increasingly includes the international and regional context. Globalization has created pressures on IR for efficiency in the employment relationship, reflected for instance on the emphasis on flexibility (types of contracts, working time, pay, etc.) and productivity. These developments and the pressures for labour market deregulation and flexibility raise the issue of efficiency versus equity.

However, the main issue for IR in this regard is not efficiency and equity as antithetic concepts, but how to achieve a balance between the two. This is because while an IR system should facilitate competitiveness, it should also promote equity by ensuring a fair return on labour and a fair sharing of the gains from economic activity, reasonable and safe working conditions, and an environment in which employees can communicate and discuss their concerns and be represented in order to protect and further their interests. According to one writer, the practices which make up equity are mainly "employee participation in employment decisions including bargaining; due process in resolving perceived injustce; security of expectations through job rights, work rules and compensation structures; and job design of a sort that is responsive to technology and organization, as well as job-holder needs."

Efficiency cannot be achieved through an IR system which is devoid of equity, particularly now when competitiveness depends so much on people, who will withhold efficiency in an environment which is inequitable and demotivating. Such withholding is often reflected in absenteeism, indiscipline, low productivity and quality, a lack of customer concern and high turnover. As such, it is efficiency and inequity which are antithetic, and it is sometimes overlooked that equity is "not an extraneous constant imposed upon the market by political institutions but rather a vital lubricant of the market process."

Further, political instability is sometimes the result of large sections of the population not being beneficiaries of economic development. This may occur when, for instance, large disparities in weal th continue to grow and there is no significant improvement in the conditions of those living below the poverty line. It has been aptly stated that:

“It is one of the least advertised, and for the very affluent the least attractive, of economic truths that a reasonably equitable distribution of income throughout the society is highly functional."

The issue of efficiency and equity arises in the debates pertaining to labour market regulation and flexibility. Employers in particular see labour market rigidities resulting from over-regulation of the labour market, especially from outside the enterprise such as fromthe plethora of labour laws in some of the South Asian countries, the orders of labour courts, and union activity outside the enterprise, all of which, according to employers, circumscribe their capacity to effect needed changes to adapt to the globalized environment.

Labour market regulation is "the creation and enforcement of rules which are designed to control the actions of individuals and groups who are a party to the production of goods and services."

In his classic work of three decades ago, Allan Flanders viewed IR as "a study of the institutions of job regulation.", through sources both internal and external to the enterprise. According to Flanders, while external sources of regulation seek to protect employees from the adverse effects of a completely unregulated market and to minimize conflicts between unions and employers, internal forms of regulation arise from employer initiatives to control employee work behaviour.

The debates about labour market regulation reflect three positions: one which favours a completely unregulated labour market; one which espouses a decentralized IR system so that external regulation is reduced, and arguments which advocate external regulation as being necessary to address a range of issues which need to be resolved for reasons of both equity and efficiency. In respect of the third argument, it has been pointed out that:

"The production of goods and services requires the coordination of activities that transform resources into an activity or product. To achieve these it is necessary to have procedures or rules that will ensure efficiency. In reality, then, the debate is not about whether there should be rules but about the source of rules and what form they should take."

If the objective is to balance efficiency wth equity, rules in a labour market system would need to be formulated from both within and outside the enterprise, the issue being one of degree or the extent of regulation.

The rules formulated from outside the enterprise may relate to basic human rights such as those enabling organizations of employees and employers to operate, rules against child and forced labour; rules to govern conflict, strikes and other work disruptions; institutions and procedures to resolve disputes; tripartite mechanisms to facilitate consensus on national labour policy; minimum terms on matters such as safet and health, minimum age of employment, holidays, maximum working hours, social security, and minimum wages where appropriate; rules against discrimination in employment e.g. on grounds of gender.

The other source of rules is the enterprise where the particular needs of both management and employees are addressed. Some writers claim that there is little evidence to suggest that an environment without some external regulation results in more efficient practices within the firm, and "research indicates that the critical factors determining good economic performance are neither internal nor external to the workplace, but come from linking both forms of regulation."Since reliance on only internal modes of regulation may well lead to rules determined solely or largely by management, the challenge for years to come will be how to achieve the balance and link between external and internal modes of regulation.

It is probably safe to predict that decentralization trends and moves to reduce external regulation are not transitory phenomena, but are likely to endure ofr many years in the foreseeable future. The fact that pressures for better corporate performance will increase, rather than decrease, is obvious. Therefore IR cannot afford to be "static" if it is to endear itself especially managements, at a time when much of management is about change. External labour market regulations which are seen as obstacles to needed change are likely to be modified. Where workers achieve higher educational levels, theymay wish to have a greater influence and voice at the workplace level over the formulation of rules in accordance with which work is to performed.

In the competition for economic superiority, market share and foreign investment, governments of fast industrializing countries are hardly likely to regulate the labour market except for the purpose of promoting efficiency and balancing it with equity. With increasing affluence, people in most societies are likely to want the fruits of development equitably shared across society. It is instructive to note that in Western Europe there has, up to now at any rate, been a deep seated commitment to balance economic achievements with social progress, so much so that education, health care, social security and quality of life have been opted for in preference to maximizing profits at their expense, thus ensuring a comfortable standard of living for most people. It is left to be seen how the evolving global economy affects the ability of such countries to maintain this emphasis. Many employers and others see Western Europe over-regulated to the point of coming close to rendering industries uncompetitive.

Freedom of Association, Labour Rights and Changing Patterns of Work

With the disappearnace of major ideological differences with the end of the cold war, unions are likely to move towards a greater concentration on their core IR functions and issues. In some Asian countries freedom of association, including labour rights in special economic zones, has rise as an issue. The need for employees and their representatives to be involved in change and in transition, and the willingness of employers to involve them, is an emerging issue in many Asian countries.

Changing patterns of work (e.g. more homework, part-time work sub-contracting) have created concerns for unions in particular. Job security, social security and minimum conditions of work are some of them. Traditional IR systems based on the concept of a full-time employee working within an enterprise is increasingly inapplicable to the many categories of people working outside the enterprise. In some countries in terms of numbers they are likely in the future to exceed those working within an enterprise. IR in the public sector - especially in the public service - where negotiation rights, for instance, are less than in the private sector, is also likely to be an issue in the future.

Women

The increasing influx of women into workforces has raised issues relating to gender discrimination, better opportunities for them in relation to training and higher-income jobs and welfare facilities.

Migration

There is a large migration of labour from labour surplus to labour shortage countries in Asia. Among the issues which have arisen are their legal or illegal status (which may affect their rights), trade union rights and their access to the same level of pay and other conditions enjoyed by nationals. Social security for migrant workers is one of the major problems as many receiving countries do not extend social security benefits to them.

Human Resource Management

With increasing reliance by employers in Asia on HRM as a means of enhancing enterprise performance and competitiveness, important consequences will arise for IR and for unions. What part unions can and will play in HRM and whether IR and HRM will operate as parallel systems (ifso what their respective roles will be), or become integrated (especially since the distinction between IR and HRM is becoming blurred), are some of the issues which will have to be addressed.

Transition Economies

In countries in transition to a market economy major challenges and issues have arisen. principally because they are seeking to adapt to an IR system in which, for instance, employers' organizations and union pluralism have hitherto been unknown. Unions in such economies may play a welfare role, and sometimes a supervisory role, rather than a negotiating tole. Managements and unions in such a system participate not so much in deciding terms and conditions of employment, but in applying decisions which are largely made outside the enterprise.

There is less scope in a centrally planned economy for tripartite dialogue between government onthe one hand and independent organizations of workers and employers on the other.

In a market economy decisions are for the most part, made within the enterprise, and where they are made externally, they are generally the result of discussions with workers' and employers' organizations representing the interests of their members vis-a-vis each other and with the government. The government creates the framework in which the social partners are consulted on matters directly affecting the interests they represent, and the social partners seek to influence the economic and social policy formulated.

Labour relations are based largely on the basis of negotiation between the two social partners, and the outcomes are usually recognized by the State so long as they do not conflict with national laws or with fundamental national policy.

Another reason for the critical role of IR in an economy in transition is the absence or inefficiency generally, during the process of transition, of safeguard mechanisms (such as for dispute prevention and settlement) at the national, industry and enterprise levels, to channel differences and disputes into peaceful means of resolution. The disputes therefore can involve considerable work disruptions and sour the environment needed to achieve sound industrial relations, and thereby also retard the achievement of overall development objectives.

In these circumstances countries in transition to a market economy are addressing a range of problems such as: the role of employers' and workers' organizations; national policy formulation through a tripartite process; a labour law system relevant to the new economic environment; methods and criteria in wage determination; dispute prevention and settlement procedures and mechanisms; and managing public sector enterprises in a competitive environment.

8

Wage Theories

- Wage: Payment for labour or services to a worker, especially remuneration on an hourly, daily, or weekly basis or by the piece.
- In economic theory, wages reckoned in money are called nominal wages, as distinguished from real wages, i.e., the amount of goods and services that the money will buy. Real wages depend on the price level, as well as on the nominal or money wages.

ECONOMIC THEORIES ABOUT WAGES

- Many theories have been advanced to explain the nature of wages. The first of them was the subsistence theory of wages, also called the "iron law of wages," of which David Ricardo was one of the main exponents.

Modern Wage Theory of Wage

Wage is a price of productive labour. The marginal productive theory, indeed, provides fairly satisfactory explanation of wage determination but its main shortcoming is that it does not consider the supply aspect of labour and concentrate on demand side

- The modern theory of wages is an extension of this theory in more logical and rational way.
- Here, the wage rate, is determined by the interaction of the forces of demand for and supply of labour in given market situation.

The demand for Labour

- Productivity of labour
- Technology
- Demand for the product.
- The price of capital input.

Shift in demand for labour

- The industries demand as a whole represent the market demand for labour.
- An industry is collection of firm.

The Reasons or Factors Affecting wage Differentials

Wage differ in different employments or occupations, industries and localities, and also between persons in the same employment or grade. One therefore comes across such terms as occupational wage differentials. Wage differentials have been classified into three categories: First, the differentials that can be attributed to imperfections in the employment markets, such as the limited knowledge of workers in regard to alternative job opportunities available else where, obstacles to geographical, occupational or inter-firm mobility of workers, or time lags in the adjustments of resource distribution and changes in the scope and structure of economic activities.

Examples of such wage differentials are inter-industry, inter firm and geographical or inter-area wage differentials. Second, the wage differentials which originate in social values and prejudices and which are deeper and more persistent than economic factors. Wage differentials by sex, age, status or ethnic origin belong to this category. Third, occupational wage differentials, which would exist even if employment markets were perfect and social prejudices were absent.

Wage differential arises because of the following factors:-

- Difference in the efficiency of the labour, which may be due to inborn quality, education and conditions under which work may be done.
- The existence of non-competing group due to difficulties in the way of the mobility of labour from low paid to high paid employments.
- Difference in the agreeableness or social esteem of employment.
- Differences in the nature of employment and occupations.

Salary Theorem

Everyone knows the Salary Theorem establishes that engineers and scientists can NEVER EVER earn as much money as businessmen, salesmen, politicians, and actors easily make. This theorem can be demostrated by reducing it to a simple mathematical equation:

The equation rests on two postulates:

Postulate N°1: Knowledge is Power

Postulate N°2: Time is Money

Given that: Power = Work/Time

And because: Knowledge = Power/Time = Money

Therefore we have: Knowledge = Work/Money

We can now easily obtain: Money = Work/Knowledge

So when Knowledge goes towards zero, Money goes towards infinity, regardless of the value attributed to work, even if the value of work is very small. On the contrary, when Knowledge goes towards Infinity, Money goes towards Zero, even if the value of Work is High. *The evident conclusion:* The less you know, the more money you definitely make. Those of you who have had difficulity following this presentation must make a Lot of Money !

Salary And Wage Management

In this article we are discussing the ways and means of fixing the compensations levels taking into consideration various factors like experience, identical industry knowledge, qualifications and proven skills. This in short we have comprehended as 'Salary and Wage Management'.

Acquire Qualified Personnel

Compensation needs to be high enough to attract applicants. Pay levels must respond to supply and demand of workers in the labour market since employers complete for workers. Premium wages are sometimes needed to attract applicants who are already working for others.

Retain Present Employees

Employees may quit when compensation levels are not competitive resulting in higher turnover of manpower. Ensure Equity: Compensation management strives for internal and external equity. Internal equity requires that pay be related to the relative worth of jobs, so that similar jobs get similar pay. External equity means paying workers what comparable workers at other firms in the labour market pay.

Reward Desired Behaviour

Pay should reinforce desired behaviours and act as an incentive for those behaviours to occur in the future. Effective compensation plans reward performance, loyalty experience, responsibilities, and other behaviours.

Control Costs

A rational compensation system helps the organization obtain and retain workers at a reasonable cost. Without effective compensation management, workers could be over or under paid.

Facilitate Understanding

System should be easily understood by human resource specialists, operating managers, and employees. Wage and salary programmes should be designed to be managed efficiently, making optimal use of the human resources information system, although this objective should be a secondary consideration compared with other objectives.

Systems to Achieve the Objectives

The above mentioned objectives are achieved by the use of the following systems.

- Job Evaluation: All jobs will be analyzed and graded to establish the pattern of internal relationships. It is the process of determining relative worth of jobs. It includes selecting suitable job evaluation techniques, classifying jobs into various categories and determining relative value of jobs in various categories.

- Wage and Salary Ranges: Overall salary range for all the jobs in an organization is arranged. Each job grade will be assigned a salary range. These individual salary ranges will be fitted into an overall range.
- Wage and Salary Adjustments: Overall salary grades of the organization may be adjusted based on the data and information collected about the salary levels of similar organizations. Individual salary level may also be adjusted based on the performance of the individual employees.

Principles of Wage and Salary Administration

There are several principles of wage and salary plans, policies and practices. The important among them are:

- Wage and salary plans and policies should be sufficiently flexible.
- Job evaluation must be done scientifically.
- Wage and salary administration plans must always be consistent with overall organizational plans and programmes.
- Wage and salary administration plans and programmes should be in conformity with the social and economic objectives of the country like attainment of equality in income distribution and controlling inflationary trends.
- Wage and salary administration plans and programmes should be responsive to the changing local and national conditions.
- These plans should simplify and expedite other administrative processes.

The Elements of Wage and Salary System

Wage and salary system should have relationship with the performance, satisfaction and attainment of goals of an individual. The following elements of wage and salary system are identified by experts in HR,

- Identifying the available salary opportunities, their costs, estimating the worth of its members of these salary opportunities and communicating them to employees.
- Relating salary to needs and goals.
- Developing quality, quantity and time standards relating to work and goals.
- Determining the effort necessary to achieve standards.
- Measuring the actual performance.
- Comparing the performance with the salary received.
- Measuring the job satisfaction gained by the employees.

- Evaluating the unsatisfied wants and unreached goals of the employees.
- Finding out the dissatisfaction arising from unfulfilled needs and unattained goals.
- Adjusting the salary levels accordingly with a view to enabling the employees to reach goals not attained and fulfills the unfulfilled needs.

Bibliography

T. Telsang Mertand: *Industrial Engineering and Production Management*, S. Chand Publisher, 2002.

Shashi Kant Yadav: *Textbook Of Industrial Engineering*, Discovery Publishing, 2011.

C Venkatramaiah and A Krishna Sharma: *A Compendium of Objective Questions in Civil Engineering*, Universities Press, 2014.

Mukund Narayan, Satyendra Kumar and Nilesh Biwalkar: *A Reference Manual of Soil and Water : Conservation Engineering*, Biotech Books, 2014.

S.N. Sarbadhikari: *A Short Introduction to Biomedical Engineering*, Universities Press, 2006.

John F. James:*A Students Guide to Fourier Transforms : With Applications in Physics and Engineering*, Cambridge University Press, 2005.

J.P. Gupta: *T B of Engineering Mech. and Basic Civil Engineering*, S. Chand Publishing, 2011.

P. C. Sharma:*A Text Book Of Production Engineering*, S. Chand Publisher , 2010.

I. J. Nagrath and Madan Gopal:*A Textbook Of Control Systems Engineering*, New Age International , 2010.

Trymbaka Murthy:*A Textbook of Elements of Mechanical Engineering*, I.K. International, 2006.

Durga Nath Dhar, Shalin Kumar and Jyoti Rai:*A Textbook of Engineering Chemistry*, Vayu Education of India, 2010.

S. S. Dara:*A Textbook of Engineering Chemistry*, S. Chand Publisher , 2010.

R. K. Dhawan:*A Textbook Of Engineering Drawing*, S. Chand Publisher , 2007.

A. Rashid : *Introduction to Genetic Engineering of Crop Plants*, I.K. International Publication, Delhi, 2009.

A.R. Dabholkar : *General Plant Breeding*, Concept Publication, Delhi, 2006.

Anil Kumar Srivastava : *Genetic Engineering and Biotechnology*, Swastik Publication,

Delhi, 2011.

Bipin Kumar Panda : *Encyclopaedia of Biotechnology and Genetic Engineering*, Anmol Publication, Delhi, 2008.

C.B. Singh : *Encyclopaedia of Plant Breeding*, Anmol Publication, Delhi, 2010.

C.P. Malik : *Genetic Engineering : A New Hope for Crop Production and Improvement*, Aavishkar Publication, Delhi, 2010.

Dinesh Arora : *Biotech's Dictionary of Genetic Engineering*, Biotech Books, Delhi, 2004.

Don Grierson : *Plant Genetic Engineering*, Springer Publication, Delhi, 2007.

E. Thro : *Genetic Engineering : Shaping the Material of Life*, Universities Press, Delhi, 2000.

E.B. Babcock : *Genetics and Plant Breeding*, Agrobios Publication, Delhi, 2007.

Grierson, Springer : *Plant Genetic Engineering*, Paperback, Delhi, 1998.

Gyan Deep Singh : *Genetic Engineering of Plants*, Anmol Publication, Delhi, 2008.

Iqbal Hussain : *Fundamentals of Plant Breeding*, Oxford Book Company, Delhi, 2009.

J.B. Jain and Sumit Jain : *Biotech's Dictionary of Plant Breeding and Genetics*, Biotech, Delhi, 2005.

James A.S. Watson, Wattie J. West and C.V. Dadd : *Crops: Varieties* Bibliography 291

and Plant Breeding, Satish Serial Publication, Delhi, 2005.

Jane K. Setlow : *Genetic Engineering : Principles and Methods*, Springer Publication, Delhi, 2010.

K.V. Mohanan : *Essentials of Plant Breeding*, PHI Learning, Delhi, 1998.

K. Venugopal and V. Prabhu Raja:*A Textbook Of Engineering Graphics*, New Age Publications , 2009.

D.A. Hindoliya:*A Textbook of Engineering Graphics*, BS Publications/BSP Books, 2014.

H. S. Gangwar and Prabhakar Gupta:*A Textbook Of Engineering Mathematics II*, New Age International , 2010.

S. S. Bhavikatti:*A Textbook Of Engineering Mechanics*, New Age Publications , 2010.

Index